MARINE WEATHER of Western Washington

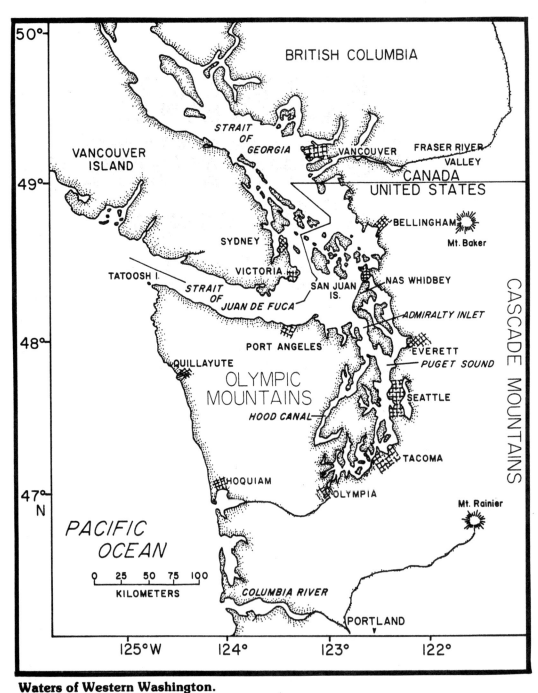

Waters of Western Washington.
Map courtesy of Pacific Marine Environmental Laboratory, Seattle, Washington.

MARINE WEATHER of Western Washington

Kenneth E. Lilly, Jr.
Commander, NOAA

STARPATH
Seattle, Washington

Copyright © 1983 by Kenneth E. Lilly, Jr.

All rights reserved. No part of this book or its appendicies may be reproduced or transmitted in any form or by any means, electronic or mechanical, including photocopying, recording, or any information storage or retrieval system, without permission in writing from the publisher.

Library of Congress Catalog Card Number: 83-50478
ISBN 0-916682-38-2 (pbk.)

Published by Starpath School of Navigation
2101 North 34th Street, Seattle, WA 98103

Manufactured in the United States of America

Library of Congress Cataloging in Publication Data

Lilly, Kenneth E.   1944 −
    Marine Weather of Western Washington

    Bibliography: p.
    1. Meteorology, Maritime -- Washington (State)
I. Title
QC994.6.L54   1983   551.69164'32   83-50478
ISBN 0-916682-38-2 (pbk.)

CONTENTS

                                                                    page

Preface viii

Chapter 1.   WEATHER PRINCIPLES

    Weather Maps 1
        Barometric Pressure 1
        Winds Around Highs and Lows 2
        Pressure Gradient and Wind 3
        Pressure Tendency 4
        Lows and Highs 4
    Weather Fronts (Warm, Cold, and Occluded) 5
        Winds and Fronts 7
        Weather and Fronts 9
        Arctic Front 12
        Stationary Front 14
    Summary 14

Chapter 2.   WAVES AND CURRENTS

    Factors Affecting Height of Waves 15
        Wind and Height of Waves 15
        Air Temperature and Waves 17
        Rainfall and Waves 17
        Tidal Currents and Waves 18
        Swells in the Strait of Juan de Fuca 20
    Ocean Wind Waves 21
    Ocean Swell 24
    Waves at Harbor Entrances 27
    Destructive Ocean Waves 28

Chapter 3.   CLIMATOLOGY

    Using Climatic Data 31
        Winter Season (December, January, February) 31
        Spring Season (March, April, May) 34
        Summer Season (June, July, August) 34
        Fall Season (September, October, November) 34
        Average Wind Directions in Western Washington 34
        Average Precipitation in Western Washington 34
    Weather Statistics for the Washington Coast 35

Chapter 4.   LOCAL WEATHER

    Convergence Zone 37
    Wind Channeling 38
    Gale Force Winds with Frontal Passage 41
    Thunderstorms 42
    Fog 44
    Sea Breeze in the Strait of Juan de Fuca 46
    The Ediz Hook Eddy 49
    Hazardous East and Northeast Winds 52

Chapter 5.   COASTAL WEATHER
    Strong Winds with Lows 55
    Super-Storms 59
    Weather Fronts 62
    Strong Winds with Highs 68
    Fog Over the Ocean 70

Chapter 6.   SAILBOAT TACTICS AND WEATHER
    Small Scale Windflow Principles 78
        Air Stability and Wind Gusts 79
        Wind Around Islands 82
        Channeling and Wind In Passes 83
        Slope Winds 83
        Winds Near Thunderstorms 85
    Local Wind Effects in Puget Sound of Interest to Sailors 86
    Ocean Sailing 88
    Buys Ballot's Law 91
    High Seas Weather Information 92

**Chapter 7.   FORECASTS, WARNINGS,
AND SOURCES OF WEATHER INFORMATION**
    Purposes of Forecasts 94
    Using Forecasts 94
        Small Craft Advisory 96
        Gale Warning 96
        Flag and Light Displays for Warnings 97
        Storm Warning 99
        Marine Forecast Areas 99
    VHF-FM Continuous Weather Radio 99
        Weather Radio KHB-60 101
        Weather Radio KIH-36 103
        Weather Radio KEC-91 105
        Weatheradio CFA-240 107
        Weather Radio WXM-62 108
    Television and Radio 109
    Telephone Recording 110
    Newspapers 110
    Radiotelephone Stations 110
    High Seas Weather Transmissions 110
    FAA Radio Transmissions 112
    Vessel Traffic Service 113

Chapter 8.   PUTTING IT ALL TOGETHER
    Using the Weather Summary 114
        Troughs 114
        Ridges 115
    Example of Using Forecasts, Summaries, and Observations 115
    Weather Check List 120

APPENDICES

Appendix A	Glossary	123
Appendix B	Wind and Wave Descriptions	126
	Beaufort Wind Scale	127
Appendix C	Wave Height Tables	
	Inland Waters and Strait of Juan de Fuca	128
Appendix D	Wave Height Tables	
	Open Ocean	138
Appendix E	Water Temperatures in Northwest Washington	141
Appendix F	Hypothermia	142
Appendix G	Wind Chill	144
Appendix H	Conversion Tables	
	Pressure: Inches to millibars	145
	Temperature: °F to °C	146
Appendix I	Cloud Identification	147
Appendix J	Additional Reading	150

Preface

Weather is unavoidable. It is one item we can never leave on the beach when we venture forth on the water, whether it be in a canoe or super tanker. Not surprisingly, the size of our craft will usually determine what weather and sea conditions are important to us, but along with these, how we use weather signs in the sky, forecasts, tidal currents and other information, along with our knowledge about our skills and equipment, may make the difference between a pleasant trip and a harrowing one.

This book provides detailed information about marine weather of western Washington. It is primarily intended for recreational and professional mariners who operate small craft (vessels less than 65 feet long), but the basic principles, examples, and sources of information are pertinent to vessels of any size in these waters. Only those situations affecting safety and comfort are discussed at length, since fair weather poses little risk.

What weather conditions are particularly hazardous? How can we recognize adverse weather situations in advance? How can we use weather observations transmitted on VHF Weather Radio to advantage? How high can the waves get? How do we make decisions using weather information available to us? These are only a few of the questions that I hope this book will help answer.

When I started investigating the weather and the effects it has on mariners in western Washington, it soon became apparent that scientists have actually made very few detailed studies of the local marine weather. The waterways in this part of the state present an extremely complex area for small scale weather and wind phenomena to occur, and the number of weather observing sites within the region is too small for detailed studies of every location. Further complicating the picture was the fact that none of the sites are actually located over the water; many are shore and inland stations, and a few are island stations.

What I felt was needed was a publication explaining the important <u>known</u> weather effects in enough detail that the diligent mariner could recognize hazardous weather situations and effectively use the weather information available to him. But some compromise on the level of detail is required. It would have been quite impossible to have done a detailed analysis of the weather and winds in every cove, bay, inlet, and pass, simply because of the lack of data. This was especially a problem along the coast, for this an area with very few nearshore observing sites.

To prepare this book, local weather studies done over the years by meteorologists at the National Weather Service Forecast Office in Seattle were examined and the pertinent information was extracted. Many of the phenomena, however, were not described in detail. Where needed, these details were filled in by my own research.

This book covers the Inland Waters of western Washington, the Strait of Juan de Fuca, and the waters along the Washington coast. The Inland Waters extend from Olympia northward through Admiralty Inlet, including the waters east of Whidbey Island, then to the area north of Whidbey Island to the Canadian border. The Strait of Juan de Fuca extends from Whidbey Island to Cape Flattery, and the coastal waters covered extend from Cape Flattery to Cape Disappointment. Each of these areas has unique winds and weather.

Throughout the book you will find numerous examples of weather events that have occurred in western Washington, and though no two weather situations are exactly the same, situations similar to these examples will certainly happen again. The examples illustrate very clearly the magnitudes of the forces that can occur. I have tried to limit the amount of weather theory to that which is essential to understanding the examples, yet the book is not oversimplified to the point where details are lost in a whitewash of broad generalities.

In short, this book is a compilation and analysis of marine weather for recreational boaters, commercial fishermen, and other professional mariners in the waters of western Washington. I hope that it might help you understand the weather you already know of, and provide those that are newer to the waters of western Washington with a source of "local knowledge" of the weather that might otherwise take years to acquire by direct experience. In either case, I believe you will find this book useful in making decisions for your operations on the beautiful waters of western Washington.

Many people gladly assisted me in preparing this publication, and to them I am gratefully indebted. The need for a marine weather guide for the small craft operator was first pointed out to me in 1971 by Mr. Stanley Marczewski, a forecaster at the National Weather Service Forecast Office in Seattle. Chapter 3 was prepared using information from an earlier climatic study of his.

Numerous people made valuable suggestions and reviewed the manuscript: Mr. L. W. Snellman, Chief, Scientific Services Division, NWS Western Region Headquarters; Mr. Donald Olsen, Port Meteorological Officer, NWS Seattle; Lt. Alan Yanaway and Lt. Mark Koehn, NOAA, marine forecasters for NWS Seattle; and Mr. Kent Short, Oceanographer, NWS Seattle. Rear Admiral E. D. Stanley, Jr., USN (ret.) and Cdr. C. L. Gott, USN (ret.), of the SEA USE Council in Seattle, were especially helpful in their suggestions and encouragement. The majority of the figures in the text were drawn by Mrs. Louise Koehn.

Finally, a special thanks to the Starpath School of Navigation and my publisher, David Burch, who suggested many improvements to the original text. Much of Chapter 6 is a result of his ideas. Information on specific local wind effects were provided by Bruce Hedrick, racing columnist and sailor with extensive experience in the Puget Sound Area.

Seattle, Washington
July 1983

Kenneth E. Lilly, Jr.
Commander, NOAA

Chapter 1
WEATHER PRINCIPLES

Familiarity with some of the concepts used by meteorologists is necessary if you want to obtain the most you can from forecasts and weather observations. Because air is a gas in constant motion, it not only moves parallel to the ground, but also up and down causing it to cool by expansion and to heat by compression. The earth's surface also cools or heats the air. Additionally, water vapor changes from an invisible gas to visible liquid (and vice versa) under various conditions, which not only forms clouds and precipitation, but also affects the temperature of the air. It is the interaction among all these motions and processes that causes our weather.

The more extreme the differences in temperature and water vapor content between large volumes of air, the more extreme the wind and weather will be. Over the past 100 years or so, meteorologists have struggled with the problems of representing three dimensional weather patterns on flat maps so they could get an overall picture of the weather situation. It has been necessary to study vertical and horizontal slices of data through the atmosphere to arrive at a complete picture of the pattern. We will only concern ourselves with one of these slices, the surface weather map. This is the map we have all seen at one time or another in newspapers and on television.

Weather Maps

A weather map is simply a drawing giving the outline of weather patterns based on observations taken at simultaneous times over a geographical area. Surface weather maps, which show the patterns next to the earth, are produced every three hours by the National Weather Service's National Meteorological Center near Washington, D.C., for use by forecasters. Other maps showing weather patterns from about 5,000 feet to over 40,000 feet are also produced and are used by meteorologists to make forecasts. The lines on the simplified surface map (Fig. 1-1) that outline high pressure areas (anticyclones) and low pressure areas (cyclones, lows, depressions) are called isobars. Isobars are lines connecting points of equal sea level pressure as determined from barometers.

Barometric Pressure

Barometric pressure is a measure of the weight of the column of air from the barometer to the top of the atmosphere. Barometric pressure is expressed in inches of mercury or millibars (mb). The conversion factor is 0.02953 inch of mercury = 1 mb. A conversion table is in Appendix H. When you hear a report that the pressure is 29.53 inches (1000 mb), it means that a column of mercury in a tube having a vacuum will stand 29.53 inches high. The higher you go in elevation, the lower the pressure will be, and so corrections are made to all barometric readings before they are put on the weather map to correct for the effects of elevation. For example, pressures at Yakima are corrected to sea level values so that meaningful comparisons with pressures at Seattle or any other location can be made. A falling or rising barometric reading often indicates a change in the weather, but not always. Pressure is especially important because a difference in pressure between two points produces a wind.

Winds Around Highs and Lows

Winds, in a very general way, blow from areas of high pressure toward areas of low pressure. However, due to the rotation of the earth and friction between the earth and the air, winds blow in a direction over the open ocean as shown in Fig. 1-2. The winds in the Northern Hemisphere, as viewed from space, blow clockwise around a high and counterclockwise around a low. The meteorologist

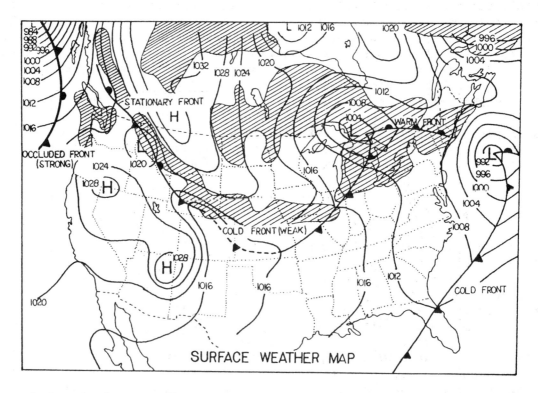

Figure 1-1 Surface weather map. Isobars labeled in millibars. An "H" denotes an area of high pressure and an "L" denotes low pressure. Hatched areas are areas of precipitation. The closer the isobars are together, the stronger the winds. Winds in the Gulf of Alaska and the Atlantic would be 30 knots or more. Actual weather station reports have been deleted to enhance clarity. Map is for 4:00 a.m. PST January 15, 1977.

uses a simple method of representing wind speed and direction (Fig. 1-3). Unfortunately, matters are not quite this simple in areas such as northwestern Washington where mountains and channels greatly affect both the direction and speed of the wind.

In our area, then, we can expect the winds to blow from high to low pressure in directions dictated by the terrain, and, in fact, this is what we often observe.

Pressure Gradient and Wind

The greater the <u>difference</u> in pressure between two points, the stronger the wind will be. This difference in pressure over horizontal distance is termed the pressure gradient (Fig. 1-4). Note that in both cases the pressure at Bellingham is higher than at Quillayute, which means the wind will be easterly in the Strait of Juan de Fuca, but will theoretically be twice as fast in (B) as compared to (A) because the pressure gradient is twice as great. This simple relationship is complicated by the vertical stability of the air, which is beyond the scope of this book. Suffice it to say that the more turbulent the layer of air over the water, the stronger the winds will be. A similar example using pressures at Portland, Olympia, Seattle, and Bellingham could be made for north-south winds in the Inland Waters. If pressures are higher at the stations south of Bellingham, the winds will be southerly in the Inland Waters, but may be easterly or westerly in the Strait of Juan de Fuca depending on whether or not the pressure at Bellingham is higher or lower than at Quillay-

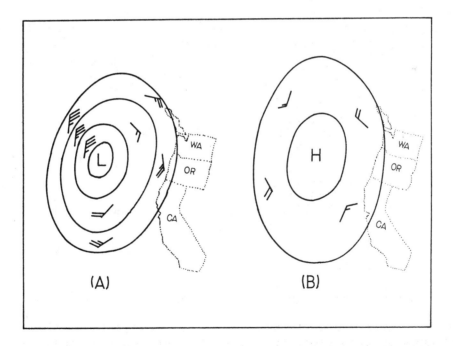

Figure 1-2 Winds around lows and highs. (A) Winds blow counterclockwise around a low in the Northern Hemisphere. Winds spiral in toward the center of the low. (B) Winds blow clockwise around a high in the Northern Hemisphere and spiral away from the center.

ute. If pressure differences between stations are 1 mb or less, the winds may not conform to the pressure gradient principle and may be light and variable (less than 5 kn. and from any direction).

Pressure Tendency

Other items you might see on the weather map are temperature, type of weather affecting the station, barometric tendency (amount of increase or decrease in the pressure over the past 3 hours), and the location of weather fronts. Pressure tendency is an important indicator of changes in the weather in many cases. Generally, the faster the change, the more intense the winds are likely to be. The change in pressure may be due to the passage of a low, the approach and passage of weather fronts or, in the cases of pressure rises, may be due to the buildup of a high over the area. A rapid rise or fall in pressure means a change in pressure of at least 6 mb or 0.18 inch of mercury in 3 hours.

Lows and Highs

Low pressure systems are usually associated with precipitation and wind and just plain lousy weather. The reason for the precipitation lies in the fact that air is forced to rise in a low due to new air from outside the low spiraling in toward the center, forcing the air upward which causes it to cool by expansion and condense the water vapor into clouds. Of course there are many variations to this simple explanation, but the principle remains the same.

Now take the case of a high pressure system. Because the air spirals away from the center, air from high altitudes must sink to replace that which is leaving the anticyclone at lower levels. The net result of the sinking motion

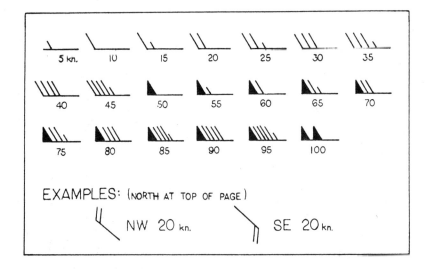

Figure 1-3 Wind speed symbols. The main shaft for the barbs "flies" with the wind and indicates the TRUE direction from which the wind is blowing. A calm or zero knot observation is denoted by a small circle with a larger circle around it. A one or two knot wind is denoted by a single shaft with no barbs.

is to heat the air by compression at some elevation above the ground, which tends to evaporate the clouds as the water droplets change back to an invisible gas. It is possible to have precipitation in a high if the earth heats up enough and there is enough moisture in the air to condense as the heated bubbles of air rise to higher altitudes. Fluffy clouds called cumulus are often associated with fair weather and high pressure systems, but these can grow into thunderheads if conditions are right.

Weather Fronts (warm, cold, occluded)

It is the boundary areas between highs and lows that cause our most violent weather in northwestern Washington because it is across these regions that the largest pressure gradients exist and the greatest changes in other properties

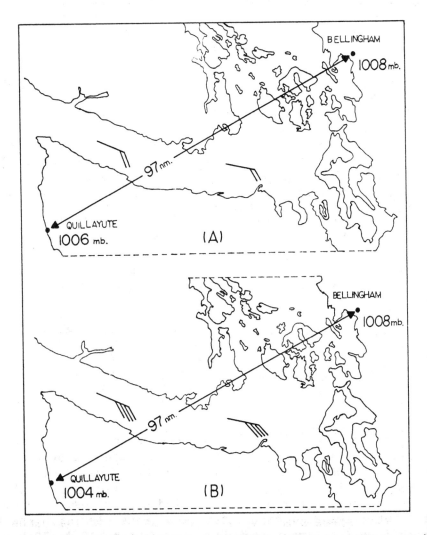

Figure 1-4 The pressure gradient principle. The pressure gradient in (B) is double that in (A). Gradient in (B) is 4 mb/97 nm or 0.0412 mb/nm and in (A) is 2 mb/97 nm or 0.0206 mb/nm.

of the atmosphere, such as temperature and water vapor content, occur. Many times we find a weather front -- or even two or three of them -- within the low pressure system itself or in the region between the low and high pressure centers.

A weather front is the boundary zone between two large masses of air having different densities on either side of the front because one mass is colder than the other. There are three types of fronts that affect western Washington: warm, cold, and occluded.

When warmer air is overtaking colder air on a large scale because of movements of large masses of air, the narrow zone separating the two masses is called a warm front. Conversely, when a cold air mass is replacing a warmer one, this narrow zone is a cold front. Many times the cold front will overtake the warm front before reaching Washington, resulting in an occluded front in which the warm front is forced aloft by the denser air behind the cold front. Fronts are often areas of high winds and stormy weather. Figure 1-5 shows how a frontal system may proceed across the eastern Pacific. As you can see, it all depends on where you are as to what type of front will affect you. There is usually a distinct shift in the wind direction from one side of the front to the other.

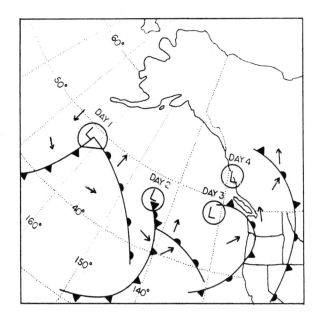

Figure 1-5 Typical movement of a low pressure center and frontal systems across the eastern Pacific Ocean. Warm front is denoted by a heavy line and half circles; cold front by a heavy line and triangles; occluded front by alternating half circles and triangles on side of line toward which the front is moving. Small arrows show prevailing wind directions. System of isobars around each low center has been deleted to enhance clarity. Fronts may move at 30 to 60 knots over the open ocean, but slow down to 10 to 30 knots over land. It is common for a front to leave its parent low center and move out ahead of it.

Winds and Fronts

The amount of change in the wind is very much controlled by the pressure gradient and topography over which the front passes. Over the open ocean hundreds of miles offshore, where the effects of land do not need to be considered, the wind is usually southeast to south ahead of the front and southwest to northwest after it passes us. Wind speeds usually increase as the front approaches us, reach a maximum as the actual front arrives, then perhaps slacken off for a few minutes only to pick up again behind the front, but from a different direction.

Location of nearby land, however, greatly affects the winds. For those waters south of Admiralty Inlet in northern Puget Sound, the change in wind direction will not be as pronounced because the winds are channeled by the terrain. Such is not the case in the Strait of Juan de Fuca where winds are usually easterly before the front arrives and westerly after it passes. Table 1-1 indicates typical conditions associated with various fronts, and Fig. 1-6 indicates the general wind direction ahead of and behind a front.

Sometimes fronts stall over our area and may extend all the way to Hawaii, bringing us rainy weather for days and days; at other times, fronts may approach us from almost due west, which causes periods of bad weather interspersed with fairly good, but cool weather as one front after another rides through the state. Fronts may also hit us from the northwest and bring snow during the winter months. Unfortunately, it is very difficult, perhaps impos-

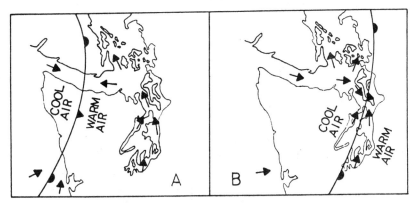

Figure 1-6 Progression of a front across western Washington. An occluded front is shown in the figure and arrows indicate wind directions. It should be mentioned that winds north of Whidbey Island may be southerly or northerly depending on the location of the low pressure center with which the front is associated. In this example, the lowest pressure is north of the area shown in the figure. Panel (A): Winds in the Strait of Juan de Fuca are easterly ahead of the front and westerly behind it. Panel (B): Winds become westerly in the Strait of Juan de Fuca after frontal passage. This is the usual case because the low center with which the front is associated moves inland north of Washington State shortly after the front passes through the Strait of Juan de Fuca. There is a chance that the winds will remain easterly after frontal passage if the low presssure remains offshore. In this case, easterly winds after frontal passage are much less than before frontal passage. Use of weather observations on VHF Weather Radio is helpful in detecting the location of a front.

TABLE 1-1 TYPICAL CONDITIONS WITH MOVING FRONTS[a]

WARM FRONTS

	Strait of Juan de Fuca		Inland Waters	
	Approaching	After Passing	Approaching	After Passing
PRESSURE	falling	rising, then falling	falling	rising, then falling
WIND	easterly	westerly[b]	S to SE	S to SW
TEMP.	rising	rising more rapidly	rising	rising more rapidly
PRECIP.	drizzle, foggy, rain increasing	drizzle, foggy	drizzle, foggy, rain increasing	drizzle, foggy
VISIBILITY	deteriorating	poor	deteriorating	poor

COLD AND OCCLUDED FRONTS

	Strait of Juan de Fuca		Inland Waters	
	Approaching	After Passing	Approaching	After Passing
PRESSURE	falling	rising	falling	rising
WIND	easterly	westerly, gusty[b]	S to SE	S to SW, gusty
TEMP.	rising or little change	decreasing	rising or little change	decreasing
PRECIP.[c]	rain, heavy showers	some clearing showery	rain, heavy showers	some clearing,[d] showery
VISIBILITY	poor in showers	improving	poor in showers	improving

[a] If fronts become stationary or very slow moving (less than 5 kt.), precipitation may continue for days with little change in any of the other conditions.
[b] Winds may remain easterly after frontal passage if the low pressure area remains offshore instead of moving inland with the front. Easterly winds will be light.
[c] Precipitation may be snow or hail instead of rain.
[d] Convergence zone (area of slack winds and usually heavy precipitation) may form in Inland Waters south of Admiralty Inlet after cold or occluded frontal passages.

sible, to forecast how strong the winds will be unless you have access to much more information than is available on VHF-FM Weather Radio or from popular media sources.

Rather than predicting the winds yourself, you are likely to have better luck using the marine forecasts. Cold fronts and newly formed occluded fronts present the greatest hazards, especially in the October to May period when the fronts are likely to be strongest. Warm fronts are usually not as hazardous with regard to wind, but often bring very poor visibilities and continuous precipitation. There are signs you can use to detect approaching fronts -- signs that have been used by mariners and farmers for centuries. Although not totally reliable, they are often harbingers of weather to come, perhaps as much as 24 hours in advance.

Weather and Fronts

Before a warm or occluded front arrives, there is typically a stream of high clouds made of ice crystals extending several hundred or more miles ahead of the front. As the moon or sun shines through the ice crystals in the cirrus (pronounced "sear-us") clouds, the light rays are bent in such a way as to form a halo around the moon or sun. This halo usually measures 22° of arc, the angle being measured from the center of the moon or sun to the inner edge of the ring.

Later on, as the winds high in the atmosphere pick up speed and strike the Olympic and Cascade mountain ranges, the air stream is forced into a wave-like pattern on the leeward side of the mountain peaks. Little patches of cloud may form in the crests of these waves and are often shaped like lenses or almonds with very distinct edges. These wave clouds may also cap off higher mountain peaks, such as Mt. Rainier and Mt. Baker, and usually appear stationary in the sky. By this time, middle level clouds at altitudes from 6,500 to 23,000 feet are invading the sky.

Called altostratus, these grayish colored clouds can easily be differentiated from the lower level clouds called stratus. It you can still see the mountain peaks above 6,500 feet, these grayish flat clouds are altostratus. When the moon or sun shines through them, an indistinct wreath of white light forms, which is called a corona. But unlike the halo, the corona will be very close to the luminous body. In some circumstances, if the clouds are not too thick, the corona can have color to it, just as the halo does. Shortly after corona formation the clouds thicken and the sun or moon gradually dwindles away to a bright spot of light which soon becomes completely obscured.

At this point, light rain may begin falling (or snow if it is cold enough). The cloud bases fall lower and lower, and the precipitation becomes steadier and heavier. Southerly winds increase over the Inland Waters, and easterly winds freshen over the Strait of Juan de Fuca. If it is a warm front approaching us, we can expect heavy rain clouds, called nimbostratus, along with other low clouds of the stratus type (stratus meaning layered or extending horizontally over large areas).

These clouds hang very low on the mountains, often with bases less than 2,000 feet up. The winds shift in direction to the southwest in the Inland Waters and to the west in the Strait of Juan de Fuca as the front passes over us. Depending on the moisture content in the air and the stability of the air behind the warm front, we may or may not get some clearing. Drizzle and fog may continue until the cold front arrives, which may be from less than an hour to over eight hours later depending how close the cold front is on the heels of the warm front.

The cold front usually has more wind and stormier weather than the warm front. Sometimes squalls of thunderheads form ahead of a strong cold front in a murderous line packing storm force winds and heavy precipitation. This squall line may travel ahead of the front by over 100 miles. Fortunately, northwestern Washington is seldom hit by violent squall lines, but they can occur and with little warning because they develop so rapidly. As the actual cold front gets closer, winds again pick up from the south in the Inland Waters and the east in the Strait of Juan de Fuca.

With a cold front, events take place more quickly. You may or may not see a sun or moon halo, but you are likely to see altocumulus clouds extending in rows at heights above 6,500 feet. Towering cumulus clouds or even thunderheads rapidly approach with heavy showers of rain, sleet, snow pellets, or hail. Winds may be well over 30 knots and gusty as the front closes in and passes. Again, as with the warm front, the wind direction shifts after passage of the front, and scattered heavy showers may persist a day or two in the cold air behind the front. Figure 1-7 shows the sequence of events associated with warm and cold fronts. A west to east profile of the front and topography was taken,

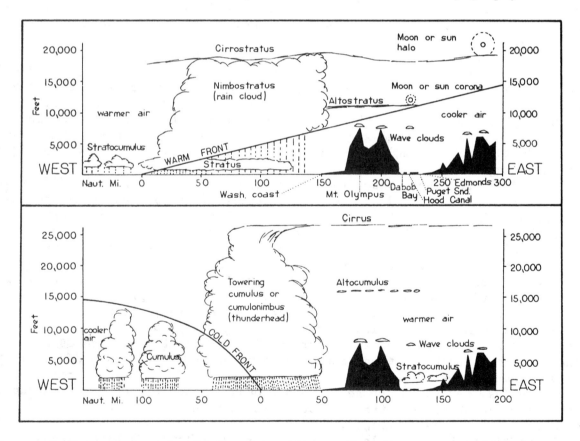

Figure 1-7 Warm and cold front profiles. Top figure: profile of warm front. Bottom figure: profile of cold front. In both diagrams, the vertical scale is exaggerated 30 times compared to horizontal scale. (For cloud identification, see Appendix I.)

but any slice perpendicular to the front as drawn on a surface weather map would reveal a similar picture.

The most common type of front to pass through western Washington is the occluded front, which has elements of the warm and cold front mixed together. Temperatures behind the occluded front may be lower or higher than the temperatures ahead of the front. If the air is colder behind the front, then it is a cold occluded front; and if the air is warmer behind the front, it is a warm occluded front. Figure 1-8 shows the two types in profile. An occluded front approaches the coast with the same signs of a warm front in many cases and may pack quite a punch if it is coming from a deep low. "Deep" is a rather subjective term used to describe the central pressure of a low and the winds associated with it. When the central pressure is less than about 975 mb or 28.80 inches of mercury, the term "deep" is often used by meteorologists. Winds around the low over the open ocean will be at least gale force and very likely storm force. Generally, the deeper the low, the more violent the weather.

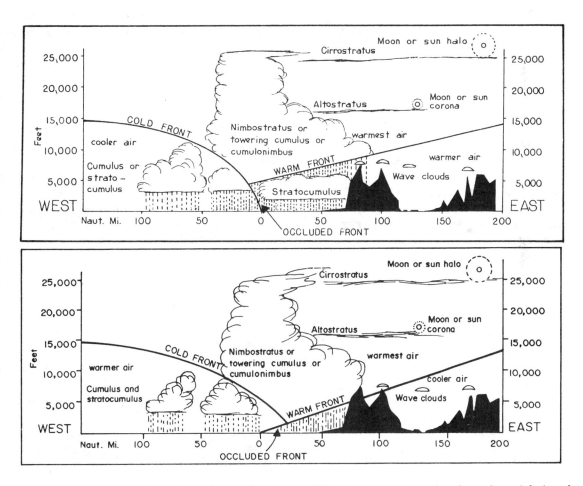

Figure 1-8 Occluded front profiles. (Top): Cold occlusion in which air behind the occluded front is colder than the air the front is displacing. (Bottom): Warm occlusion in which the air behind the occluded front is warmer than that ahead of the front. The fronts in both panels are assumed to be moving from west to east, and the vertical scale is exaggerated 30 times compared to the horizontal scale. (For cloud identification, see Appendix I).

There is more energy in a deep low so its fronts will be stronger and with higher winds.

If the occluded front is weak, the weather and winds will be more like those associated with a warm front, but if the occluded front comes from a deep low, it will have effects similar to a cold front. The formation of occluded fronts is a fairly complex matter and is beyond the scope of this book. Suffice it to say that one reason they form is due to the cold front overtaking the warm front because it travels faster. However, since the advent of weather satellites in the 1960s, meteorologists have become aware of numerous cases where occluded fronts form almost instantly without going through the overtaking process. These are called instant occlusions.

Instant occlusions are often very dangerous to boating because they can form so rapidly in a deepening low that may only be a few hundred miles off the Washington coast. Marine warnings may only give a short notice of several hours.

You can learn a lot about clouds and how to read the sky through the use of well-illustrated books of cloud pictures and being very observant of the sky. A few books having excellent cloud pictures along with explanations of their significance are listed in Appendix J.

Arctic Front

Although there are only three basic types of fronts, there are some variations of these basic types that affect western Washington: the arctic front and the stationary front. The arctic front, also known as a cold air outbreak, is a special brand of cold front, but instead of coming from the Pacific Ocean it comes from the north, out of Canada. It occurs only in winter or early spring since it depends on extremely cold air bottled up over British Columbia for its main source of energy. Figures 1-9 and 1-10 show how the outbreak is triggered.

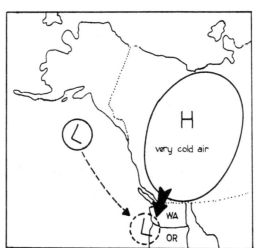

Figure 1-9   Principle behind the cold air or arctic outbreak. If a strong high over 1030 millibars (30.42 in.) builds over British Columbia, a low moving in close to the Washington coast may trigger the outbreak. The low may come from almost any direction over the Pacific.

As a high pressure system settles over the snow-covered interior of British Columbia, the air mass under the high gets colder and colder, building up a pressure head in the valleys and passes to the interior of the province. The high may attain pressures over 1040 mb (30.71 inches of mercury). If a low develops over the Gulf of Alaska or anywhere in the eastern Pacific Ocean and then moves toward Washington, a tremendous pressure gradient is set up across the valleys and passes, and the wind comes whistling down the Fraser River Valley north of Bellingham in the form of a strong cold front.

Winds behind this front will often be 30 knots or higher from a northerly direction in the Inland Waters and from an easterly direction in the Strait of Juan de Fuca. Temperatures will usually be less than 32°F. As the front sweeps southward through Puget Sound, steep and rough seas are created in areas

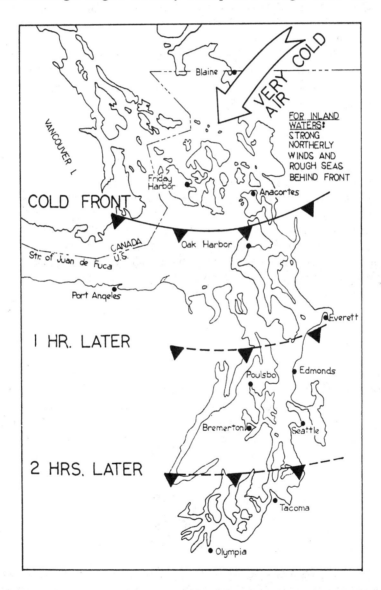

Figure 1-10 Progression of arctic front or cold air outbreak through the Puget Sound area. Strong winds and high wind chill may continue for several days.

unprotected by land masses. The author observed the effects this had at the Anacortes Pier one spring when high waves made it impossible for a 175-foot ship to dock. The wind chill effects will be very high, and the wind may persist for several days. Sometimes heavy snow showers occur with the front followed by clearing behind it.

Ninety percent of arctic air outbreaks occur between December and February with January being the favored month. Not every year has a major outbreak, but some years have over four. The very cold air has been known to stay over our area up to three weeks, but an average of three days is more common depending on the prevailing weather pattern. Temperatures of 0°F to -8°F may occur during major outbreaks in western Washington, especially if the outbreak is accompanied by a snowfall. But as the weather pattern changes and the outbreak weakens its frigid grip, another weather front from the Pacific is likely to encounter this very cold air, which may result in a heavy snowfall followed by much warmer temperatures. Eventually, all the cold air is scoured out, and the weather turns to rain.

Stationary Front

The other variation of frontal activity, the stationary front, is simply a front that has stopped moving. This brings frustrating weather because the precipitation seems to continue on and on with very few sun breaks. If the stationary front results from a stalled front that came from the north in the winter or early spring, we have extended periods of snow or rain and snow mixed. If the stationary front results from a front that came from the Pacific Ocean, we get long periods of rain and warmer temperatures. The forecaster watches these fronts very carefully because small storm disturbances may develop along them, which increases the wind and precipitation as the disturbances ripple into western Washington. The timing of the arrival of these disturbances along the main front is often very difficult.

Summary

The most violent type of weather is associated with strong pressure gradients caused by fronts (especially cold and occluded fronts), depressions or lows, and the rapid buildup of high pressure into our area after a storm has passed. Rain or snow may force us into the comfort of our cabin (or make us wish we had one), but it is the wind, waves, and tidal currents that affect our safety the most. The effects and interplay of winds and waves and currents are the topics of Chapter 2.

Chapter 2
WAVES AND CURRENTS

Factors Affecting Height of Waves

When a wind blows across a smooth body of water, small ripplets appear almost immediately, then wavelets and chop, and finally, more fully developed waves having more definite wave lengths and shapes. Figure 2-1 depicts some of the basic facts about waves and their development. Wave length is the distance between two crests or two troughs. The height of the wave is the distance between the trough and crest. Small ripplets and chop have heights less than 1 foot and very short wave lengths so that a small power boat gives the feeling that it is traveling on a road of washboards when it is planing. Winds of about 1 to 6 knots will create these small waves. With increasing winds, however, more energy is quickly transferred to the water, and the waves continue to develop until they reach the maximum height attainable under those particular conditions. The following factors affect the height of waves:

1. Wind speed
2. Wind duration
3. Length of wind fetch (distance of water over which the wind blows)
4. Width of the body of water
5. Depth of water
6. Set (direction) and drift (speed) of water currents
7. Air temperature
8. Rate of rainfall

Wind and Height of Waves

Wind speed, duration, and fetch are the most important factors in creating waves for a given body of water. For a given wind speed, waves on the open ocean get much higher than those which develop on the waters of western Washington. In addition, wave development on the open ocean will continue for a longer period of time before the maximum wave growth is achieved. This is not to say that waves dangerous to small craft are not created on the Inland Waters, for they certainly are. We should always keep in mind that the larger the body of water is and the closer we get to the Pacific Ocean the greater the risk is that we will encounter much higher waves than in the Inland Waters. Especially in the Strait of Juan de Fuca, we should be aware of expected changes in wind direction -- a 30-knot westerly wind behind a front is going to create higher waves than a 30-knot easterly wind ahead of the front since the fetch for a westerly wind is longer. There is also the possibility of ocean swell coming from the west which would make matters even worse.

The stronger the wind, the more quickly the waves develop, and the higher they will become. However, because wave development takes place on the Inland Waters and Strait of Juan de Fuca in areas bounded by land, the wave height is restricted by both the fetch length and width. Another complicating factor is that for a certain wind speed over a given fetch, the maximum wave development will occur in a limited time period (the time period increases with increasing fetch) after which the same wind will not create higher waves no matter how long it blows. An example will illustrate these points.

Let us consider the effects of a 40-knot southerly wind blowing one hour over Lake Union in Seattle and Puget Sound off Edmonds. The fetch length and width, as well as depth of water, in Lake Union are considerably less than Puget Sound, so right away we know that waves will be much smaller in Lake Union than on the Sound. The limited fetch in Lake Union will also result in an increase in wave height with distance from the southern shore. The north end of the lake will have larger waves. This effect is also prominent on Lake Washington near the floating bridges. Even in light winds you can notice the wave height increase with distance from the leeward side of the bridges.

Waves developing in water less than about 100 feet deep will also be smaller than those in deeper waters. Since Puget Sound is deeper than 100 feet and Lake Union is less than 100 feet, the waves will be higher on the Sound. After one hour, the 40-knot wind will create waves of about 1.5 feet at the north end of Lake Union and about 4.8 feet off Edmonds. In the Sound an occasional highest wave of about 9 feet may occur. Wave height tables for various points in the Puget Sound area are given in Appendix C.

These tables take into account wind speed, wind duration, length of wind fetch, width of fetch, and depth of water so you will not have to do long calculations to arrive at an estimate of wave heights. The heights in the tables are what are termed "significant wave heights," which are an average of the heights of the highest one-third of the waves. These values are used because they most nearly represent the heights that would be reported by an experienced mariner. Maximum wave heights of the occasional highest waves may be almost double those shown in the tables. Figure 2-2 and Table 2-1 illustrate the significant-wave-height definition.

In comparison to the waves on the open ocean, waves on the Inland Waters and Strait of Juan de Fuca are quite puny. If we took the 40-knot wind in our

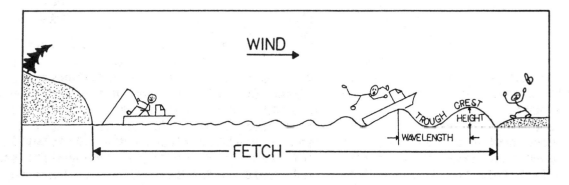

Figure 2-1   The greater the fetch, the higher the waves.

example above and applied it over a fetch length of at least 100 nm in the open ocean, we would end up with 6-foot waves in one hour and 16-foot seas in six hours. Waves of over 80 feet have been recorded in Queen Charlotte Sound of British Columbia! These high waves were, of course, caused by winds much higher than 40 knots.

Waves often times look much higher than they actually are when observed from a small boat because you are so close to them. Winds of less than 15 knots will usually not cause too much difficulty for small craft in most areas, but even this wind can make uncomfortable conditions in large bodies of water, such as the Strait of Juan de Fuca.

Air Temperature and Waves

Cold air at 32°F is denser than air at 70°F, and denser air has more force on the water as it blows across it. For a given fetch area and wind, the waves will be higher in a cold wind than in a warm one. How much higher is difficult to pinpoint, but it is on the order of 10 to 15 percent. Cold winds behind a strong cold front or an arctic air outbreak will create higher waves than a comparable wind having higher temperatures.

Rainfall and Waves

Almost all mariners know that a thin layer of oil on the water reduces wave development. A similar effect occurs when a heavy downpour of fresh water covers the more dense sea water. It restricts the development of waves as the less dense liquid tends to slide with the wind instead of forming waves. The dampening effect is very noticeable in tropical downpours. Far more important than air temperature or precipitation, however, is the effect tidal currents have on wave development.

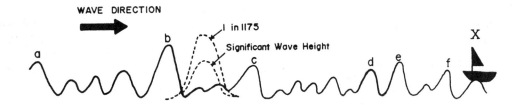

Figure 2-2 Wave train profile. Vessel at point X experiences a wide variety of wave heights as the waves move from left to right past his position. Of the 18 waves shown here, waves a-f represent the highest one-third and would be averaged to determine the significant wave height. (In actual practice, a much larger number of waves would be used.) The significant wave height will be higher than waves a, c, d, and f but less than wave b. Statistically, one wave in 1175 will be 90% greater than this value. The unusually high wave and the significant wave are shown as dotted lines. Wave lengths in the figure are not to scale.

TABLE 2-1. WAVE HEIGHTS FROM SIGNIFICANT WAVE HEIGHTS

Most frequent wave height:	0.5 x SWH
Average wave height:	0.6 x SWH
Significant wave height (average height of highest 33%):	1.0 x SWH
Height of highest 10% of the waves:	1.3 x SWH
Height of highest 1% of the waves:	1.7 x SWH
One wave in 1175 waves:	1.9 x SWH
One wave in 300,000 waves:	2.5 x SWH

Tidal Currents and Waves

Currents going with the waves decreases their heights while currents going against the waves (or within about 45° of the opposite direction) cause them to grow higher and steeper. How much higher, to the author's knowledge, has never been measured in the Puget Sound area, but studies in open ocean regions having currents over three knots indicate wave heights may be up to twice as high as compared to the same waves in an area where there are currents of one knot or less. The steepness of the face of the wave is more important that the height of the wave (Fig. 2-3) because a boat no longer rides easily over the wave, but gets slammed instead. Very steep and breaking waves are extremely hazardous to small craft.

There are a number of areas in the Inland Waters where ebb and flood currents exceed one knot, but there are only a few places where average currents regularly exceed three knots. These locations are shown in Fig. 2-4 -- other areas may have currents over 3 knots, but only irregularly. If winds are going to be over 20 knots in a direction opposing a current of three knots or more, you should make allowances for higher and steeper waves. Several U.S. Government publications put out by the National Ocean Service of the National Oceanic and Atmospheric Administration are helpful in locating areas where currents are strong and for computing their velocities.

<u>United States Coast Pilot 7: Pacific Coast, California, Oregon, Washington, and Hawaii.</u> (Describes in detail many areas of the Inland Waters and Strait of Juan de Fuca with regard to navigational hazards, such as rip currents.)

<u>Tidal Current Tables - Pacific Coast of North America and Asia - 19XX</u>
(This set of tables is published annually and gives daily predicted currents at hundreds of different locations.)

<u>Tidal Current Charts - Puget Sound Southern Part and Tidal Current Charts - Puget Sound Northern Part</u> (These two charts depict the ebb and flood current patterns at various stages of the tide.)

<u>Nautical Chart 18445 - Puget Sound, Possession Sound to Olympia</u>

Including Hood Canal, Washington and Nautical Chart 18423 - Bellingham to Everett Including San Juan Islands, Washington (These two sets of charts depict in detail numerous items of interest, such as navigational aids, current patterns, weather warning displays, etc.)

If you want an extremely detailed picture of how the water flows at various stages of the tide in the Inland Waters south of Fidalgo Island (located just north of Whidbey Island), the following publication may be useful:

Figure 2-3 Tidal currents and wave steepness. Top panel: Waves are not steep in an area of slack current or current going in the same direction as the waves. Bottom panel: Higher and steeper waves occur in an area when the current opposes the direction of wave travel. What were previously fairly good conditions can become hazardous. Harbor bar entrances along the coast where rivers meet the sea and places in the Inland Waters where strong tidal currents occur are likely areas for steep waves.

Lincoln, John H. and Noel McGary. <u>Tide Prints - Surface Tidal Currents in Puget Sound</u>. (Washington Sea Grant Publication WSG 77-1). Division of Marine Resources/University of Washington, Seattle, 1977.

The data in the booklet were derived from photographs taken of flow patterns in the water using a scale model of the Puget Sound basin.

Swells in the Strait of Juan de Fuca

Swell is made up of waves that develop on the open ocean during a storm and then later travel out of the areas of generation. They may travel thousands of

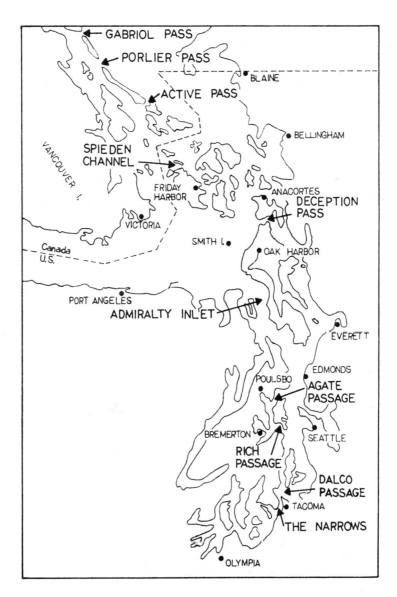

Figure 2-4 Locations where currents <u>regularly</u> exceed 3 knots. Strong winds blowing against these currents may create hazardous conditions for small craft.

miles to become a crashing surf on a distant beach. It is possible to have a swell in the Strait of Juan de Fuca, which, when combined with gale force winds from the west, may create hazardous seas in the Strait west of Whidbey Island. The author knows of one instance where a tug and barge were forced ashore along the west side of Whidbey Island under these conditions. Fortunately, these are pretty rare occurrences, but it is precisely the unusual events that cause us the most trouble.

Ocean Wind Waves

Although much is understood about the creation and propagation of ocean waves, it is still beyond science to predict where and when "killer" waves will occur, which is one reason why we should know our limitations and those of our vessel before we venture out on the open sea. If we have only experienced waves on the Inland Waters, even under gale wind conditions, we will be unpleasantly surprised by the much greater ocean waves under similar winds.

Mentioned earlier was the fact that waves on the open ocean will grow much higher in the same period of time under a given wind as compared to the Inland Waters. Figure 2-5 shows a comparison of rate of wave development off Edmonds and off the coast where the fetch is much greater. In both cases, the wind is southerly at 40 knots. We will assume the fetch to be 300 nm over the ocean

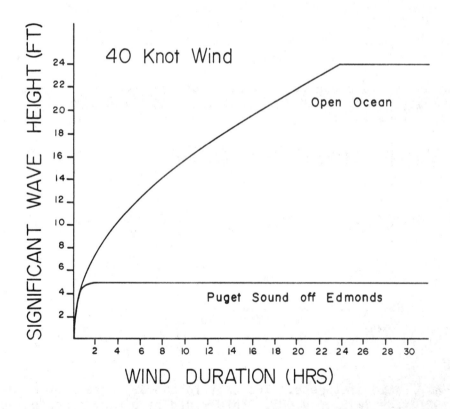

Figure 2-5 Wind wave development over the open ocean versus a mid-channel point in Puget Sound off Edmonds assuming slack water in the Sound. The wind fetch is assumed to be 300 nm on the ocean and 5 nm off Edmonds.

(fetches on the ocean may range from about 100 nm to well over 1000 nm). The effective fetch (incorporates effects of width of body of water) off Edmonds is 5 nm. Wave heights over the ocean are 10 ft. at the end of four hours compared to only about 5 feet off Edmonds. The wave tables in the Appendix should help you in making decisions when it comes to winds and waves.

The next thing to bear in mind is that the waves will usually be steeper early on during a strong wind, and the steepness is often more important than the height of the waves. Some mention of wave steepness was made earlier with regard to more protected waters. Let us see how the concept applies to ocean waves.

Wave steepness is used to describe the relative slope of the face of the waves by using the ratio of the height of the wave to its wave length. The higher the wave is for a certain wave length, the steeper the wave is. The steepness ratio gives us a means of comparing relative steepness between waves, but it should be pointed out that the true slope of the wave face is often much greater than the steepness ratio. This is because the shape of a wave is irregular with the downwind side steeper than the upwind side. Figure 2-6 shows a couple of idealized wave profiles in which the wave steepness varies. In the figure, the wave heights are the same, but the wave lengths are different.

Theoretically, all waves break when the steepness ratio is 1/7. Real waves, however, start curling over and breaking when the ratio is only 1/10. They will appear menacing and steep at a ratio of 1/18 or so.

Wave steepness changes through the course of a storm and is illustrated from an actual case in Figure 2-7. The data are from an environmental data buoy anchored in the Pacific southwest of the Columbia River entrance approximately seven miles offshore. This particular example also shows the effect of terrain on local winds. When the atmospheric pressure is higher in eastern Oregon and Washington than it is along the coast, winds will blow down the Columbia River Gorge because they are funneled by the high bluffs and mountains on either side of the gorge. Because it takes time for waves to build up, it is not

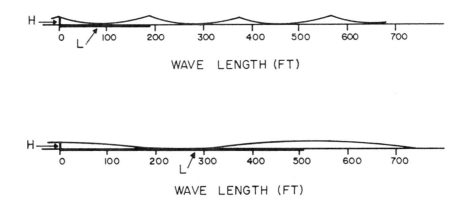

Figure 2-6 Wave steepness. Top profile shows a steep wave, having a ratio of 16 ft./192 ft. = 1/12 or 0.083. This would be a very threatening wave to small craft and would have a very menacing appearance because the wave is nearly breaking. Bottom profile shows a wave that is not steep, the ratio being 16 ft./512 ft. = 1/32 or 0.031. This would be more typical of swell.

surprising that maximum wave heights are not achieved until after the peak wind. The time lag may vary from less than an hour to over six hours depending on the weather pattern.

We can determine the wave height by estimation. Remember, waves will look much larger than they actually are from a small vessel, yet will often be

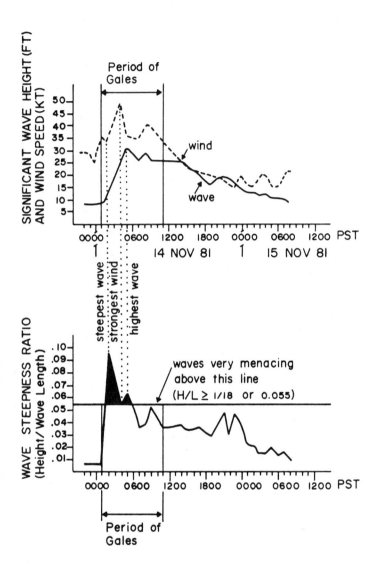

Figure 2-7 Wind speed and wave height versus wave steepness from the ocean buoy anchored off the mouth of the Columbia River. On 14 Nov. the wind was blowing easterly out of the Columbia River gorge at 25-30 knots prior to 1:00 a.m. The predominant waves were swells with eight foot heights and 0.008 steepness. As the deep low center swept up the coast, winds switched to SE-SW at gale to storm force until 11:00 a.m. on the 14th. Note that the waves reach their greatest height (31 ft.) after the maximum wind (50 kt.) but reach their greatest steepness (0.096) prior to the maximum wind. This series of events is often typical in a storm. The winds again became easterly after 2:00 a.m. on the 15th.

underestimated from a large ship. If you know your eye elevation above the water line, waves will be this height when you observe them to be even with your eye when your vessel is on even keel in a trough. It becomes much more difficult to estimate wave heights when the waves tower over you. Practicing on land by estimating heights of structures (i.e., one, two, three story buildings and so on) will help you get the feel for how tall something appears. Sailors have their mast or spreader heights as references. To get the other part of the formula for figuring out wave steepness, we need to get an estimate of the wave length.

This is best done by an indirect method since it is even more difficult to estimate horizontal distances than vertical ones on the open sea. A rough estimate of the wave length is calculated by measuring the period of the waves in seconds and using one of the formulas we will shortly come to. Period, of course, is just the time in seconds it takes two successive crests (or troughs) to pass a fixed point. By observing the up and down motion of flotsam or foam patches, it is possible to time the passage of crests (or troughs). It is best to take about ten of these measurements and average them to arrive at the period. The formulas for wave length are:

$$\text{WAVE LENGTH (ft)} = \begin{cases} 3 \times P^2(\text{sec}) & \text{for wind waves} \\ 5 \times P^2(\text{sec}) & \text{for swells.} \end{cases}$$

These formulas are simplifications of ones found in H. O. Publication 603, <u>Practical Methods for Observing and Forecasting Ocean Waves by Means of Wave Spectra and Statistics</u>.

As an example, let us assume the wind is blowing very strongly, and whipping up seas to 16 ft. We determine the wave period to be eight seconds. Using the first of the two formulas, a wave length of 3x8x8 = 192 feet is computed, and the wave steepness would be 16/192 or 1/12. This, indeed, is a steep wave. If we were observing swells, the period for the 16 ft. waves might be 12 seconds. This results in a 5x12x12 = 720 ft. wave length and wave steepness of 1/45 -- a very gentle wave by comparison.

You will know when you are in steep waves because your vessel responds accordingly. Table 2-2 lists tactile clues that describe wave steepness according to appearance of the waves and the action they have on vessels about 35 to around 65 ft. in length.

Ocean Swell

Even when there is no wind and the sea is smooth as a mirror, there almost always is a slight movement of a vessel due to swell. Figure 2-8 shows strong wind areas associated with a distant storm off the West Coast. The wind waves travel out of the fetch area at individual speeds of about 18 knots to over 48 knots. (An estimate of wave speed in water deeper than about 100 fathoms can be determined by multiplying 3 times the wave period in seconds. For example, an 11-sec. wave will have a deep water speed of 3x11 = 33 knots.)

The wind waves leaving the head of the fetch actually travel outward in an ever widening curved arc and not just straight ahead. Wave energy pumped into

TABLE 2-2. DESCRIPTIONS OF WAVE STEEPNESS

Very Hazardous (very steep)

Face of waves looks like a wall of water. Crests overhang and break. Difficult to impossible to make headway into waves. Vessel slams easily. Danger of pitchpoling, swamping, or capsizing if waves high enough. Personnel flung about by force of waves; loose articles thrown off tables; bowl of soup nearly impossible to eat. Example: breaking waves on harbor bar; some storm waves in initial stages of storm; "rogue" wave evolving out of stormy sea.

Steep

Face of waves appears less ominous than above; vessel slams easily if too much headway into waves is made and pitches uncomfortably in following seas. Waves can cause extreme rolling and risk of capsizing, but pitchpoling not likely. Personnel have great difficulty moving about without using supports and handrails and may occasionally be flung about. Loose articles slide off tables. Soup sloshes out of bowl even when the bowl is held down. Examples: mature storm waves.

Moderate

Waves appear more like swells. Vessel does not slam much at normal cruising speed and will ride up and over waves somewhat uncomfortably, but safely. Waves do not appear threatening. No overhanging wave crests. Personnel can move about with only moderate difficulty. Loose articles slide back and forth on tables. Soup in a bowl sloshes about but little or no spillage. Examples: Waves during the 6 to 24 hours after a storm has died out.

No Problem

Waves very rounded and smooth. Vessel can proceed at full speed in almost any direction without danger and rides very comfortably. Vessel rides easily up face of waves with no slamming. Personnel can move easily about without using supports or handrails. Soup in a bowl is barely disturbed and does not slosh out. Example: long ocean swell under light to calm wind conditions.

the water moves along at one-half the speed of the individual waves traveling through this moving area of energy. Gradually, due to friction between parcels of water, the waves decrease in height and become more rounded and smooth in shape. The wind waves have turned into swell approximately 12 to 24 hours after leaving the fetch.

The waves start to feel bottom when the depth of water is equal to one-half their wave length. Thus, a wave 700 ft. long will feel bottom in depths of 350 ft. and less; a 400 ft. wave length will feel bottom in 200 ft. of water. When a wave feels bottom, the wave length starts decreasing, the height of the wave increases, and the speed of the wave slows down. Only the period remains constant. When the water depth equals 1.3 times the wave height, the wave peaks up and breaks.

Waves in shallow water (shallow being determined by the wave length and depth of water) can be reflected off vertical obstructions, such as seawalls; they can be bent around islands and headlands; and they can be focused by underwater shoals. These effects can cause wave energy to be concentrated at a spot seemingly out of proportion to the swell observed in deep water. The behavior of waves in shallow water is complicated, and the book <u>Waves and Beaches</u> by Willard Bascom will prove very interesting reading if you want more detailed information.

Seldom hazardous to small vessels on the open ocean, swell may be quite discomforting to ships that roll or pitch in sympathy with the waves. The author was dramatically introduced to this fact early in his career during his first sea assignment aboard a Coast and Geodetic hydrographic survey ship. Early one morning off the coast of southern California, we were laying to, preparing to make a Nansen cast. In this operation a dozen or so two-foot long metal cannisters are attached at intervals along a cable that is lowered over the side to collect water samples at various depths. The Nansen bottles are attached to the cable as it is lowered, which, in our case, meant that the person doing the attaching had to stand on a small platform extending out from the quarterdeck and lean over the guard chain to reach the cable. I had the watch on the bridge and another fellow officer was supervising the cast.

The 229-ft. vessel rolled easily in the moderate 8-ft. to 12-ft. swells, but just as the first bottle was entering the water, she started rolling at ever increasing angles until the inclinometer was oscillating back and forth to 45 degrees. The natural period of roll of the ship had synchronized with that of the swell. The waves pooped the quarterdeck, pinning everyone who was not able to grab a support against the superstructure, and the officer on the platform was submerged to his chest in the water. Luckily, he did not wash overboard.

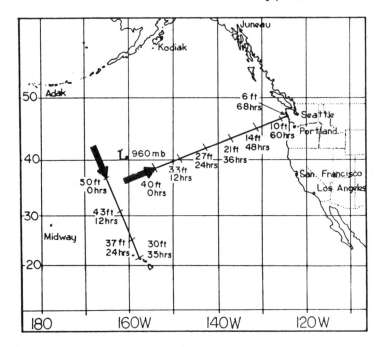

Figure 2-8 Approximate significant wave heights and travel times from a storm some 1500 nm southwest of the Washington coast. The waves approaching the coast are, in this example, initially 40 ft. high with a 15 sec. period in the fetch area where they are created. By the time they near the Washington coast some 68 hours later, they are only 6 ft. high. The 50 ft. waves with an 18 second period created to the west of the low arrive at Oahu as 30 ft. waves only 35 hours later.

From below, the crash of breaking dishes and runaway furniture reached the bridge at about the same time that the Captain called. This, I soon found out, was not the way he wanted his morning wake-up call made. Fortunately, the cast was not at depth, we got underway, and I only had to face the angry stares of the steward's department for a day as they cleaned up the mess below decks.

Waves at Harbor Entrances

All harbor mouths along the Washington and Oregon Coast, including the Columbia River entrance, can be very hazardous because of steep or breaking waves caused by shoaling and strong currents flowing against incoming waves. The height and steepness of the waves at the entrances depend primarily on the period of incoming waves, height of incoming waves, depth of water, and strength of the current. The current is usually a combination of tidal current and river current.

During periods of high river discharge and strong ebb tidal currents, conditions are set for very dangerous waves at any of the bars along the Washington coast or the mouth of the Columbia River. Anytime the ebb currents are three knots or more, watch out for steep or breaking waves if wave periods are less than 8 seconds. The shorter the wave period and the higher the incoming wave is, the less current it takes to cause the wave to break on the bar. Table 2-3 illustrates how bar conditions change with various wave characteristics on the Columbia River bar and the bar at Grays Harbor. The information

TABLE 2-3   WAVE CHARACTERISTICS ON THE COLUMBIA RIVER BAR AND GRAYS HARBOR BAR

Incoming Significant Wave Height (ft.)	Incoming Wave Period (sec.)	Ebb Current (knots)						
		1	2	3	4	5	6	
4	6	4.4	5.0	5.9M	7.0B	7.0B	7.0B	WAVE
	8	4.2	4.6	5.1	5.9	7.3M	7.7B	
	10	4.2	4.5	4.8	5.3	5.9	7.1	
6	6	6.4	7.4M	8.5B	8.5B	8.5B	8.5B	HEIGHT
	8	6.3	6.9	7.9	9.2M	11B	11B	
	10	6.1	6.5	7.1	7.9	8.9	10	
8	6	8.9M	10B	10B	10B	10B	10B	
	8	8.5	9.1	10	12M	13B	13B	
	10	8.4	8.9	9.5	10	12	14	
10	6	11M	11B	11B	11B	11B	11B	ON
	8	11	12	13M	14B	14B	14B	
	10	10	11	12	13	15	17M	
	12	10	11	12	12	13	14	BAR
12	8	13	14	15M	16B	16B	16B	
	10	13	13	14	16	18M	20B	
	12	13	13	14	15	16	18	
14	8	15	16M	17B	17B	17B	17B	
	10	15	15	16	18	20M	25B	
	12	14	15	16	17	19	21	

M -- Menacing, steep waves with ratio of height to wave length 0.055 to 0.07.
B -- Breaking waves.

Note: Wave heights on bar are rounded to nearest foot for values 10 ft. and up.

here may not apply elsewhere and is not intended to replace information you receive on actual conditions from the U.S. Coast Guard or other boats near the bar. Forecasts of bar conditions are available on VHF NOAA Weather Radio along the coast.

Table 2-3 only goes up to 14 feet for the incoming waves. In general, waves higher than this that go across the bar are likely to be hazardous in themselves because they occur with nearby storms and are not long period swell.

It is quite easy to get into serious difficulty under hazardous bar conditions, so this is one area requiring good seamanship practices. Keep in mind that when you are approaching land from seaward, the backside of the waves always appear less threatening than the front side (face of wave). Appearances are deceptive, and it is not hard to get in too close only to find out that it is also too late. Suddenly, you are being battered by very steep or breaking waves and may lose complete control of your vessel. The problem is compounded at night or under conditions of poor visibility.

The United State Coast Pilot 7 - Pacific Coast: California, Oregon, Washington, and Hawaii and the pamphlet, 13th Coast Guard District Guide to Hazardous Bars (published 1979), are two references that will give you detailed information about the bars. Not all bars have been studied in enough detail to predict their conditions, but some success has been achieved in forecasting for Grays Harbor and the Columbia River mouth. These forecasts, as mentioned earlier, are available on VHF NOAA Weather Radio along the coast.

Destructive Ocean Waves

> "At midnight I was at the tiller and suddenly noticed a line of clear sky between the south and southwest. I called to the other men that the sky was clearing, and then a moment later I realized that what I had seen was not a rift in the clouds but the white crest of an enormous wave. During twenty-six years' experience of the ocean in all its moods I had not encountered a wave so gigantic. It was a mighty upheaval of the ocean, a thing quite apart from the big white-capped seas that had been our tireless enemies for many days. I shouted, 'For God's sake, hold on! It's got us!' Then came a moment of suspense that seemed drawn out into hours. White surged the foam of the breaking surf. We were in a seething chaos of tortured water; but somehow the boat lived through it, half-full of water, sagging to the dead weight and shuddering under the blow. We baled with the energy of men fighting for life, flinging the water over the sides with every receptacle that came to our hands, and after ten minutes of uncertainty we felt the boat renew her life beneath us. She floated again and ceased to lurch drunkenly as though dazed by the attack of the sea. Earnestly we hoped that never again would we encounter such a wave."

This quotation from the 1916 diary of Sir Ernest Shakleton, written while he was crossing the ocean from Elephant Island to South Georgia in the South Atlantic, very aptly describes a rogue wave -- a wave unusually high and steep in relation to those around it. It is a wave of this type that is very destructive on the open sea and one which no seaman wants to experience. Large ships have been known to be badly damaged by such waves more than 80 feet above the water line and some have even been sunk. Again, like height of waves and steepness, a rogue wave to a small vessel may not be called that at all by a person on a large ship. It all depends on the characteristics of the wave and the size of your vessel.

Waves on the ocean are made up of a large number of different waves having various heights, periods, and wave lengths. It is not surprising to find that waves occasionally reinforce each other, creating unusually high and steep waves compared to those prevailing around the area. In a storm tossed sea, such waves can appear seemingly out of nowhere, roll ominously along for a few minutes, and then disappear. It is not possible to forecast when and where such waves will appear, but something can be said about the odds of encountering such waves based on statistics gathered on waves.

Predicted wave heights and frequencies are listed in Table 2-1. From that table you can see that in a storm with waves of significant wave height (SWH) 15 feet one would expect about 1 out of 1,175 waves to be about 30 ft. high. If the period of the significant waves were some 8 sec, we would expect this large wave every couple hours or so.

To see further how we might use this information, assume we receive a forecast in which the significant wave height is predicted to be 20 ft. Using the above information we can make a reasonable estimate of the range of wave heights we will likely encounter. The most frequently observed waves will only be 10 ft. and the average 12 ft. These, however, are not the waves reported by experienced observers as these are overshadowed by the much larger ones. The significant wave height, which is the average height of the waves typically reported by mariners as representative of the waves, is 20 ft. One in ten of the waves will average 26 ft., while one in 100 will be about 34 ft. We will likely experience some waves around 38 ft., but should be surprised if one 50 ft. passes through. It is these waves that are much greater than the significant height that are recalled as giants or rogue waves. Figure 2-9 shows the sizes of these waves compared to a 30 ft. boat. As a general rule, if the forecasted significant waves height verifies, we can reasonably expect a few waves at least 90% higher than this, but very rarely 150% higher.

The actual seas we must sail through are generally a combination of wind waves and swell. Their combined height, however, is not quite the sum of the two. Oceanographers have found that wind waves and swells combine to form a significant wave height according to the following formula:

$$\text{Combined Sea Height} = \sqrt{(\text{wind wave height})^2 + (\text{swell height})^2}.$$

If the reported SWH is 5 ft. and the reported swell is 6 ft., you should expect combined seas of some 7 or 8 ft., not 11 feet.

It is prudent seamanship to keep all forward hatches and ports dogged down when there is any doubt about waves coming over the bow. There are cases on record where vessels have been running comfortably in moderate seas or swell only to chance upon one of these waves much higher than the significant ones.

Green water was taken all the way up to the bridge in some cases. Stories written by people who have had experience either as professional mariners or sailboaters on the open ocean abound with incidents of rogue and monster waves. These make interesting reading and much can be learned from them.

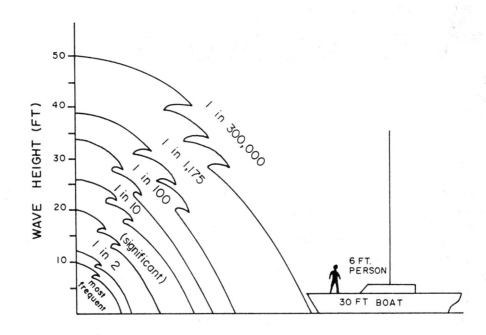

Figure 2-9   Various wave heights associated with 20 ft. significant waves as compared to a 30 ft. boat and six ft. person. Slope of waves has been exaggerated.

Chapter 3
CLIMATOLOGY

Using Climatic Data

We can learn some useful facts about the area we live in by studying climatic records because it is through these records that we get a peek at the extreme conditions found in the weather as well as the average conditions. The average conditions are determined using a weather record of 30 years in which observational values at a given site are averaged to arrive at a mean value. The average temperature for the month of March at Bellingham would be one example.

Through the use of climatology, we can make reasonable judgments for long range planning. A simple example of this would arise if we were planning to take a 60-foot vessel to Kodiak, Alaska, from Seattle -- a trip which can be very hazardous if we are not careful about the weather in the Gulf of Alaska. Examination of storm track data and frequency of gales based on years of records would reveal that the late fall to early spring months are the riskiest. Of course, day-to-day weather is usually a lot different than the average, just as the balance in our checking account each day is usually quite different than the average monthly balance. Table 3-1 shows some of the climatic values for various western Washington locations where weather observations have been made regularly over the years.

The climate of northwestern Washington on the west side of the Cascade Mountains is a middle-latitude west coast marine climate, having comparatively dry summers and usually mild and wet winters. The daily and annual range in temperatures is small compared to areas in eastern Washington. Some of the factors influencing the climate and daily weather are: prevailing direction of wind; temperature of the eastern Pacific Ocean; the Olympic and Cascade mountain ranges; amount of cloud cover; position and intensity of large high and low pressure areas over the West Coast and North Pacific Ocean; and the large expanse of waterways in northwestern Washington. Let's take a look at prevailing conditions during each of the seasons.

Winter Season (December, January, February)

Storm systems are most intense over this period. On the average, there is a permanent low pressure system hovering over a broad region off the eastern end of the Aleutian Island chain of Alaska that generates storm after storm in the North Pacific Ocean during a typical season.

TABLE 3-1 SEASONAL WINDS AND TEMPERATURES OF WESTERN WASHINGTON

Location	Average Wind Speed (knots)	Highest Sustained Wind Speed (kn)*	Average Temp. (°F)	Record Low (°F)	Record High (°F)
WINTER SEASON (December - February)					
Longview	--	--	40.8	4	73
Willapa Hbr. (South Bend)	--	--	42.0	7	72
Aberdeen	--	--	41.5	10	72
Quillayute Airport	6.1	SE 40	40.2	7	72
Tatoosh I.	16.6	S 76	43.2	14	64
Port Angeles	7.7	E 50	40.0	7	67
Port Townsend	--	--	41.0	5	67
Blaine	--	--	38.6	-1	66
Bellingham	--	--	38.2	-4	67
Olga (Orcas I.)	--	--	40.0	-8	65
Anacortes	--	--	41.1	4	65
Oak Harbor (Whidbey I.)	8.5	--	39.9	3	65
Bremerton	--	--	40.4	15	62
SEATAC Airport	8.6	SW 44	40.3	0	70
Olympia	6.4	SSW 48	39.2	-8	72
SPRING SEASON (March - May)					
Longview	--	--	49.5	19	97
Willapa Hbr. (South Bend)	--	--	49.7	20	93
Aberdeen	--	--	48.2	22	96
Quillayute Airport	5.8	SE 29	46.3	19	89
Tatoosh I.	11.8	E 79	47.6	25	81
Port Angeles	8.8	W 46	47.3	21	83
Port Townsend	--	--	49.5	19	88
Blaine	--	--	47.9	11	85
Bellingham	--	--	48.0	10	85
Olga (Orcas I.)	--	--	48.6	14	83
Anacortes	--	--	49.0	18	88
Oak Harbor (Whidbey I.)	7.1	--	48.6	16	85
Bremerton	--	--	49.4	21	89
SEATAC Airport	8.3	SW 36	49.2	11	93
Olympia	6.2	SW 40	48.5	13	92

* Sustained winds are measured over a one minute period.
-- Missing entries mean data were not avialble from these stations.

Table Data are based on these numbers of years of records:

Port Angeles (4--for average wind; 2--for sustained wind; 30--for temperatures)
Port Townsend (30--for average temperatures; 67--for record lows and highs)
SEATAC Airport (33--average wind; 15--highest sustained wind; 38--temperatures)
Olympia (27--average wind; 31--highest sustained wind; 38--temperatures)
Longview (24)   Willapa Hbr. (30)   Aberdeen (24)
Quillayute Airport (15) Tatoosh (63)   Blaine (24)
Bellingham (30)   Olga (30)   Anacortes (24)
Oak Harbor (15--for wind; 30--for temperatures) Bremerton (10)

Spring Season (March, April, May)

The Pacific Northwest begins to feel the effects of the high pressure system that _usually_ exists year around off the southern California coast. This high pressure system gradually moves northward so that weather fronts become weaker as they try to penetrate through the high. An occasional potent front passes through western Washington.

TABLE 3-1 Continued. SEASONAL WINDS AND TEMPERATURES OF WESTERN WASHINGTON

Location	Average Wind Speed (knots)	Highest Sustained Wind Speed (kn)*	Average Temp. (°F)	Record Low (°F)	Record High (°F)
SUMMER SEASON (June – August)					
Longview	—	—	62.7	37	103
Willapa Hbr. (South Bend)	—	—	60.3	36	101
Aberdeen	—	—	59.6	37	100
Quillayute Airport	5.0	SE 24	57.7	33	99
Tatoosh I.	8.7	S 63	55.1	43	88
Port Angeles	10.5	N 37	57.7	37	93
Port Townsend	—	—	60.2	37	96
Blaine	—	—	60.5	37	92
Bellingham	—	—	59.5	29	94
Olga (Orcas I.)	—	—	58.9	37	92
Anacortes	—	—	60.5	37	93
Oak Harbor (Whidbey I.)	5.7	—	59.8	35	98
Bremerton	—	—	62.4	40	99
SEATAC Airport	7.2	SW 25	62.7	38	99
Olympia	5.4	WSW 28	61.8	30	104
FALL SEASON (September – November)					
Longview	—	—	52.9	8	98
Willapa Hbr. (South Bend)	—	—	53.2	11	97
Aberdeen	—	—	52.5	11	93
Quillayute Airport	5.0	SE 37	50.5	20	92
Tatoosh I.	12.9	S 82	51.3	19	80
Port Angeles	6.4	W 41	50.0	12	85
Port Townsend	—	—	51.6	12	85
Blaine	—	—	50.0	6	88
Bellingham	—	—	49.9	3	90
Olga (Orcas I.)	—	—	50.7	10	87
Anacortes	—	—	51.6	13	85
Oak Harbor (Whidbey I.)	6.4	—	50.4	8	85
Bremerton	—	—	51.9	10	89
SEATAC Airport	7.6	S 57	52.1	6	94
Olympia	5.3	SW 52	50.8	-1	96

Data in Table 3 are based on information published by NOAA's National Climatic Data Center, Federal Building, Asheville, NC 28801. Comprehensive records are not available for most locations, but what is published appears mainly in three publications:
  "Local Climatological Data — Monthly Summary...(specify month, year, and location)",
  "Local Climatological Data — Annual Summary...(specify year and location)", and
  "Climatography of the United States — Climate of...(specify location)".
Data for Quillayute, SEATAC, and Olympia are current through 1982. Data for all other locations are based on records for various years and are not current through 1982.
The information in the table is only intended for comparisons, to show approximately how conditions vary between locations.

Summer Season (June, July, August)

By now the high pressure cell has become well developed and stops the majority of storms from reaching the coast with full force. Winds of gale force seldom occur during these months. This is the period when western Washington experiences periods of a few warm days followed by a flow of cool and cloudy coastal air into the Puget Sound region, which breaks the warm spell. Days are often cloudy and gray with some afternoon sun breaks. Sea breezes from the west are common in the Strait of Juan de Fuca. If the high moves into eastern Washington, hot easterly winds bring very warm temperatures.

Fall Season (September, October, November)

The high pressure cell that has dominated the weather in our area weakens and begins a southward retreat as more and more storms start battering it from the west. A high incidence of fog may occur in the Inland Waters and persist for days because of cool night temperatures. If the high stays over us or slightly inland, we experience "Indian Summer," which has some of the finest weather of the year. Nonetheless, by late October the dominant high has been replaced by a series of dominant lows, and the rains return.

Although these scenarios are typical in a climatic sense, they can be substantially different in an individual season. The high pressure cell did not move southward in the winter of 1976-1977, and near drought conditions were experienced in Washington because the storm systems were blocked from the coast. The positioning of these large high pressure cells is perhaps the most critical factor influencing extremes in weather.

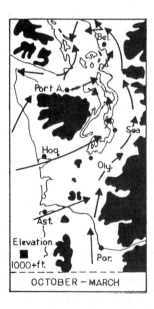
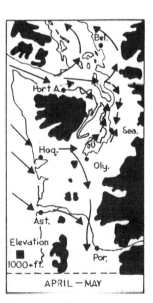
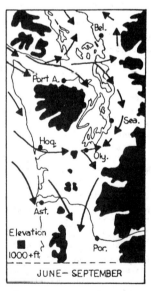

Figure 3-1 Average wind directions over western Washington. Note the predominantly southerly winds over the Puget Sound area except during the summer. Dark areas represent elevations above 1,000 feet.

Average Wind Directions in Western Washington

We close this section by taking a look at Figures 3-1 and 3-2. Figure 3-1 shows the average windflow pattern over western Washington. One thing that stands out is that winds are northerly in the Puget Sound area during the summer months, due to the high pressure system prevailing over our area, and southerly during the fall-winter months when storm systems prevail.

Average Precipitation in Western Washington

Figure 3-2 shows the average annual precipitation over western Washington. As you can see, it varies widely due to the location of mountain ranges. Precipitation is especially low in the area immediately west of Whidbey Island where a large region is in the rain shadow of the Olympic Mountains.

Weather Statistics for the Washington Coast

TABLE 3-2[a]

METEOROLOGICAL TABLE FOR COASTAL AREA OFF ASTORIA
Boundaries: Between 46°N., and 48°N., and from 127°W., eastward to coast

Weather elements	Jan.	Feb.	Mar.	Apr.	May	June	July	Aug.	Sep.	Oct.	Nov.	Dec.	Annual
Wind ≥ 34 knots (1)	6.3	5.5	2.9	3.1	*	*	0	0	*	2.6	4.8	6.0	2.6
Wave height ≥ 10 feet (1)	35.4	32.6	32.1	11.9	10.9	4.7	1.4	2.4	4.7	19.7	18.5	39.3	16.2
Visibility < 2 naut. mi. (1)	3.6	4.2	3.0	2.6	2.9	2.0	2.3	4.1	5.5	5.7	4.4	4.3	3.7
Precipitation (1)	23.6	21.2	17.9	12.6	10.5	8.7	8.2	7.2	8.8	13.9	24.1	21.1	14.8
Temperature ≥ 85°F (1)	0	0	0	0	0	0	*	*	0	0	0	0	*
Mean Temperature (°F)	45.0	46.4	46.5	48.9	52.7	56.6	60.5	61.0	59.7	56.0	50.7	47.9	52.8
Temperature ≤ 32°F (1)	2.5	*	*	0	0	0	0	0	0	0	0	*	*
Mean relative humidity (%)	81	84	80	81	81	82	80	83	83	82	82	82	82
Sky overcast or obscured (1)	47.7	48.3	42.8	36.5	41.1	45.6	44.4	34.7	32.9	39.6	50.1	45.4	42.3
Mean cloud cover (eighths)	5.9	6.1	5.9	5.5	5.7	6.0	5.4	5.4	4.7	5.6	6.0	6.0	5.7
Mean sea-level pressure (2)	1015	1017	1015	1017	1019	1018	1019	1018	1017	1016	1016	1015	1017
Extreme max. sea-level pressure (2)	1040	1037	1037	1036	1034	1030	1034	1030	1032	1034	1037	1038	1040
Extreme min. sea-level pressure (2)	980	982	986	986	1003	994	1002	1004	995	991	988	968	968
Prevailing wind direction	S	S	S	NW	NW	NW	NW	NW	N	S	S	S	NW
Thunder and lightning (1)	0	*	0	*	*	0	0	*	*	*	*	*	*

METEOROLOGICAL TABLE FOR COASTAL AREA OFF SEATTLE
Boundaries: Between 48°N., and 50°N., and from 129°W., eastward to coast

Weather elements	Jan.	Feb.	Mar.	Apr.	May	June	July	Aug.	Sep.	Oct.	Nov.	Dec.	Annual
Wind ≥ 34 knots (1)	4.3	3.3	2.3	1.6	1.1	*	*	*	1.0	2.2	3.3	2.9	1.9
Wave height ≥ 10 feet (1)	11.1	26.0	20.4	17.6	6.3	4.6	4.5	2.4	6.4	24.6	15.0	22.1	12.6
Visibility < 2 naut. mi. (1)	5.6	4.7	3.7	1.9	2.7	3.7	6.2	8.0	6.2	6.4	6.5	4.0	5.1
Precipitation (1)	28.7	25.0	19.6	17.1	14.8	11.5	10.2	6.2	12.9	19.2	29.2	28.8	18.1
Temperature ≥ 85°F (1)	0	0	0	0	0	0	0	0	*	0	0	0	*
Mean Temperature (°F)	43.6	44.9	45.5	48.3	52.0	56.3	59.4	60.6	58.1	54.0	48.6	45.7	52.0
Temperature ≤ 32°F (1)	3.6	1.0	*	*	0	0	0	0	0	0	1.2	.8	.6
Mean relative humidity (%)	81	83	80	81	80	80	81	83	81	82	81	83	81
Sky overcast or obscured (1)	52.2	49.8	39.5	42.2	40.8	38.0	38.8	38.9	35.2	40.3	45.6	51.3	42.4
Mean cloud cover (eighths)	6.2	6.0	5.5	5.7	5.7	5.7	5.1	5.3	4.9	5.5	6.0	6.3	5.7
Mean sea-level pressure (2)	1014	1015	1015	1017	1017	1017	1019	1018	1017	1015	1016	1014	1016
Extreme max. sea-level pressure (2)	1041	1041	1039	1033	1035	1031	1034	1030	1037	1038	1041	1042	1042
Extreme min. sea-level pressure (2)	980	974	984	978	991	984	997	998	988	977	975	974	974
Prevailing wind direction	SE	S	W	NW	NW	NW	NW	NW	NW	NW	SE	S	NW
Thunder and lightning (1)	*	0	*	*	0	*	*	*	*	*	*	*	*

(1) Percentage frequency.
(2) Millibars.
* 0.0-0.5%

[a] Reprinted from the U.S. Coast Pilot Vol. 7.

These data are based upon observations made by ships in passage. Such ships tend to avoid bad weather when possible, thus biasing the data toward good weather samples.

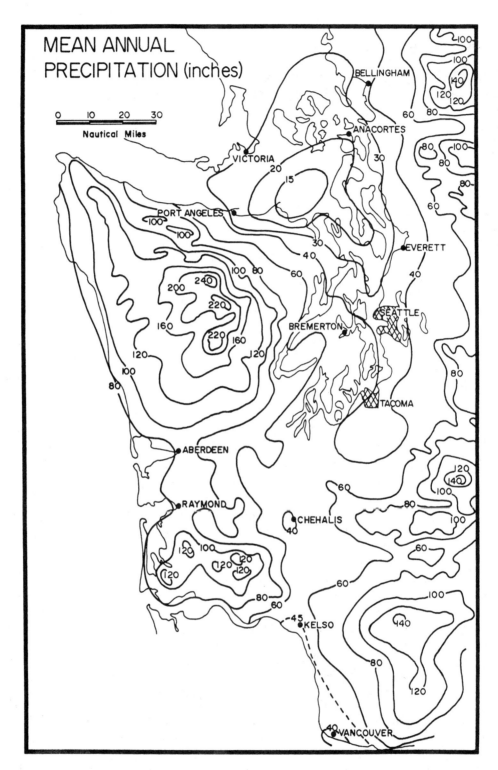

Figure 3-2 Average annual precipitation in western Washington. Note the rain shadow area east of Port Angeles.

Chapter 4
LOCAL WEATHER

Numerous local weather and wind patterns occur over the waters of western Washington, but only a few of them are critical to safety; most just affect our comfort and pleasure. We will take a look at the more important peculiarities but especially at those that may prove hazardous. The method used to illustrate the various situations employs actual weather events that have occurred plus many of the concepts discussed in chapters 2 and 3 of this book. Enough details are included so that with careful study a feel for the magnitude of wind speeds (plus wave height from the Wave Tables in the Appendix) can be obtained.

Convergence Zone

A convergence zone (CZ) is an area where opposing winds meet, resulting in vertical movement of the opposing air masses and slack winds inside the CZ. It is often the bane to sailboaters who are becalmed inside the CZ until it passes. Heavy clouds and heavy precipitation sometimes occur with the CZ, perhaps even thunderstorms if there is enough moisture in the air. At other times, little, if any, cloudiness occurs. Winds on one side of the CZ will be northerly and on the other side southerly — which means, for example, that you can be heading northward in Puget Sound on good southerly winds only to encounter head winds from the north as the CZ passes.

Most CZs initially form in the vicinity of Everett where they will stretch from the Olympic Mountains to the Cascade Range. They usually move southward at 1 to 10 knots and dissipate before reaching Tacoma. However, they also can move northward, stopping just south of Pt. Wilson; or they may simply remain stationary. Occasionally, two or three distinct CZs may form, one after the other. They are estimated to be from less than 5 miles to more than 15 miles wide and change width with time.

Especially favorable circumstances exist for the formation of CZs during winter and spring months when numerous weather fronts pass through western Washington. This is so because CZ formation depends on an onshore flow of a west to northwest wind that splits into two streams as it tries to get around the Olympic Mountains. Figure 4-1 shows the basic principle of how CZs form. They may form without a frontal passage, however, whenever an onshore wind from the ocean is of sufficient strength to be split. In Figure 4-2 are some actual cases of convergence zones. Winds are usually not hazardous with them, but heavy precipitation can ruin what otherwise was a fairly nice day. Note that the pressure gradient between SEATAC and Bellingham is very slight (less than 1.5 mb). The pressure at Quillayute must <u>initially</u> be higher than that at SEATAC or Bellingham.

Convergence zones can be detected through the use of wind observations broadcasted on VHF Weather Radio. Look for northerly winds in one part of Puget Sound and southerly winds in another part. The area in-between is the CZ. Winds in the Strait of Juan de Fuca will be westerly. The convergence zone itself is not mentioned directly in the forecasts. If westerly winds are forecasted for the Strait of Juan de Fuca or a front is forecasted to pass through western Washington, the possibility for a CZ somewhere in Puget Sound exists. In these conditions, sailboaters can expect to meet an extended area of light or calm winds as they sail up or down the Sound.

Wind Channeling

There is probably no more difficult event to forecast than what exact wind

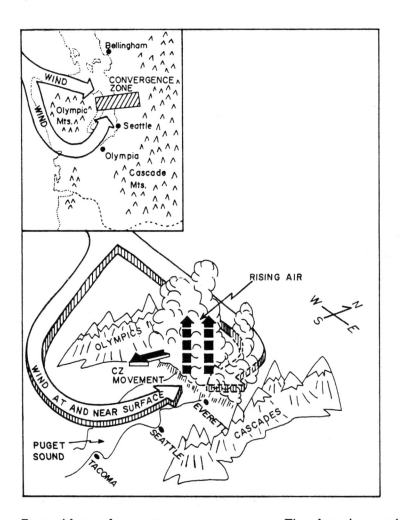

Figure 4-1 Formation of a convergence zone. The inset panel shows that the main air flow splits around the Olympic Mountains. The two streams then meet in the vicinity of Everett. Precipitation may be very heavy.

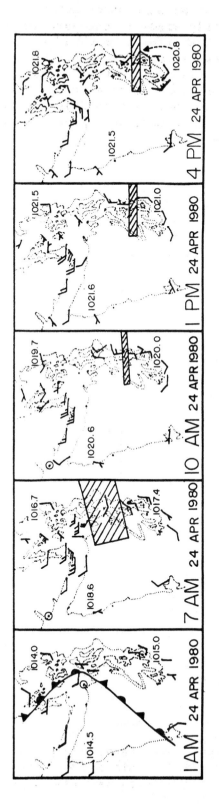
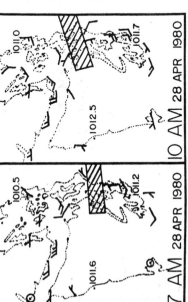

Figure 4-2 Actual convergence zones over Puget Sound. TOP PANEL: Southward moving CZ (shaded area). Convergence zone formed about 6 hours after frontal passage. Note westerly winds in Strait of Juan de Fuca. CZ dissipated by 7:00 p.m., PST. Pressures to nearest one tenth millibar shown for Quillayute, Bellingham, and SEATAC. VHF Weather Radio broadcasts only report pressures to nearest whole mb. MIDDLE PANEL: Northward moving CZ. Convergence zone dissipated by 1:00 p.m. BOTTOM PANEL: Stationary CZ. Convergence zone dissipated by 10:00 a.m.

speed and direction will occur at a given location in the Inland Waters. The general windflow determined on the large scale weather maps only gives the prevailing direction and speed, not the specific winds that occur in every channel and inlet. Winds are generally northerly or southerly in the Inland Waters and easterly or westerly in the Strait of Juan de Fuca, depending on the barometric pressures at Bellingham, Quillayute, and Olympia (using the pressure gradient principle of Chapter 2). However, there are many local effects where the general flow is disturbed by the terrain. It is possible to have strong southerly winds in most of the Inland Waters, yet have strong easterly winds through channels oriented east-west, such as Harney Channel located between Orcas and Shaw Islands in the San Juan Islands.

Figure 4-3 shows the basic principles of wind channeling. The wind blows

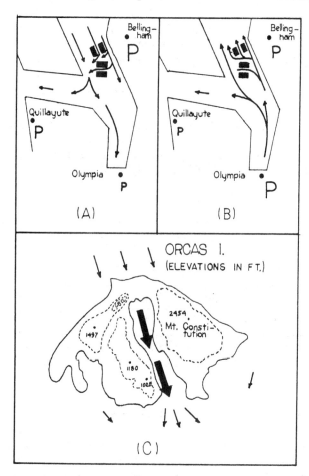

Figure 4-3 Wind channeling. The larger the "P", the higher the pressure. In (A) the pressure at Bellingham is higher than at Quillayute or Olympia so winds will flow out of the Strait of Juan de Fuca and from the northerly directions through the Inland Waters, except they may also be channeled by terrain to flow out of channels between islands. In (B) the pressure at Olympia is highest, so winds will be southerly in the Inland Waters and easterly through the Strait since the pressure at Bellingham is greater than that at Quillayute. Again, island channels may divert the main flow. In (C) very strong channeling occurs in East Sound of Orcas Island due to funneling by the mountains on both sides of the island.

from high to low pressure, is stronger the greater the pressure gradient, and can blow faster where constrictions funnel the wind. We might liken the islands, hills, and mountains to rocks in a river, the air in this case being the "water" flowing over and around the obstructions. In a river, we can see the swirls and eddies formed on the downstream sides of the rocks; the rapid increase in speed between two closely spaced rocks as the water jets through; the turbulence and split flows on the upstream sides of the rocks; and the rapid speed increase where the water makes a bend along a vertical bank. The air blowing through the Inland Waters and the Strait of Juan de Fuca behaves similarly.

It is quite possible that the forecast only calls for small craft advisories (winds 21 - 33 kn.), but the actual wind is blowing gale force out of passes at over 34 knots, such as East Sound on Orcas Island. Keep in mind that tidal currents are often strong in constricted channels and can pose an additional hazard. In seeking safe anchorages, keep in mind the location of the anchorage in relation to expected wind shifts. Neah Bay, for example, offers good shelter from westerly winds, but not easterly ones. Wind gusts are also important, especially for sailboaters. Gusts of 50 to 60 knots can easily occur within average winds of 30 knots if the air is turbulent, since turbulence can cause higher winds aloft to momentarily hit the earth's surface.

Figure 4-4 shows the basic cause of turbulent winds. Wind gusts occur when vertical swirls in the air force stronger winds higher up to move down to the earth's surface. These swirls, caused by heating of the air and obstructions to the wind, force the air stream to move up and down. Although gusts are short-lived, they are powerful. The force of a 40-knot wind is four times that of a 20-knot wind for air at the same density because the force goes up with the square of the speed. An 80 knot wind has 16 times the force of a 20 knot wind.

Light and variable winds, which are winds less than 5 knots and from virtually any direction, occur when there is less than a millibar difference in pressure between Quillayute and Bellingham or between SEATAC and Bellingham. Winds may blow from any direction and are strongly influenced by very local effects, such as heating of nearby land. At night, near shore, though the forecast is for variable winds in the region, the winds may blow steadily from the shore, for as the land cools, the air flows back out over the water.

Gale Force Winds with Frontal Passage

In Chapter 2, we looked at weather fronts in some detail because it is the wind and weather associated with fronts that are most hazardous to boating. Figure 4-5 shows the winds that occurred with a strong occluded front as it moved through western Washington. Note that the winds remained southerly in the lower Puget Sound area, but shifted from easterly to westerly in the Strait of Juan de Fuca after the front passed. The gale force winds with this front are fairly typical of the winds we find with late fall to early spring frontal systems. However, wind speeds can vary over a broad range, and so we have to rely on forecasters to give us an estimate of the winds because we simply do not have enough information available to us to make reasonably accurate predictions.

Not only can strong winds be dangerous, but also the weight of heavy snows that occur with some fronts may prove hazardous to property. In the winter of

1979-1980, over 60 boats were sunk in the Portland-Vancouver area due to heavy snows collapsing boat shed roofs. On occasion thunderstorms are generated by fronts or very unstable air in western Washington.

Thunderstorms

All of us have seen those magnificent white clouds towering like gigantic cauliflowers into the sky. Those towering cumulus and cumulonimbus clouds (thunderheads) are the most violent of the clouds. Any small boat operating under a thunderstorm is potentially in grave danger if it does not have proper lightning grounds for masts and antennas. An open cockpit runabout affords no protection from lightning.

Lightning bolts, which typically pack a punch of 125 million volts at 10,000 to 345,000 amperes (about 1,250 to 43,000 times as powerful as a nuclear power generating plant) can literally shatter the hull of a boat if it is hit.

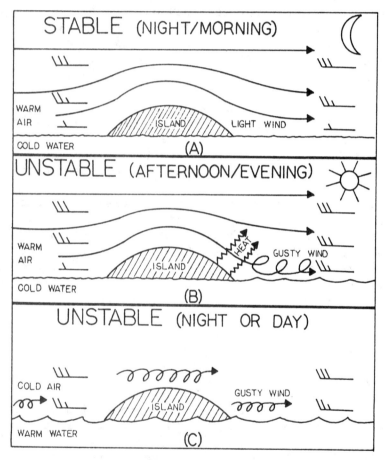

Figure 4-4   Air turbulence. (A) Cold water under warm air cools the lower air layers making them more stable. (B) By the afternoon, the sun may have heated nearby land or islands enough to cause heat bubbles in the lower air layers, which causes vertical mixing into the layers above. This can bring winds from higher up down to the water in gusts. (C) Cold air over water that is warmer is heated from below so the lower air layers mix with those higher up. Again, gusty winds are likely.

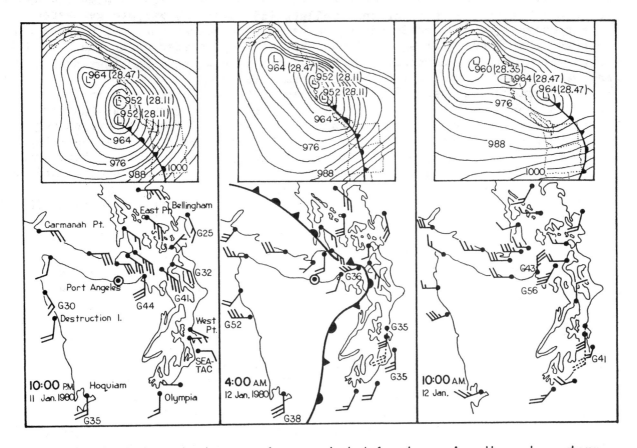

Figure 4-5 An actual case of an occluded front passing through western Washington. The insets show the large scale weather pattern with the center of each low center labeled in millibars and inches of mercury. The 952 (28.11 in. of mercury) millibar low is a very deep low. Other isobars are also labeled in millibars. Note that the isobars are closely spaced which indicates that the winds are strong. Times are PST.

At 10:00 p.m., the main low center is off Vancouver Island, and the front is approaching Washington from the southwest. Winds in the Strait of Juan de Fuca are averaging over 30 knots. Port Angeles has no wind because it is protected by the mountains. A "G" followed by a number indicates gust speed.

At 4:00 a.m., the front has already passed through most of the Strait of Juan de Fuca. Note how the front bulges near the western end of Whidbey Island. This is due to the fact that the front travels faster over water than it does land as it races eastward through the Strait of Juan de Fuca.

At 10:00 a.m., the front has almost passed through Washington. In the strong winds behind the front, swirls and eddies, much like those in a fast flowing river, have formed. East of Port Angeles at Dungeness Spit, winds are north 35 knots with gusts to 43 knots, while at Smith Island a short distance away they are southwest at 35 knots, gusting to 56 knots. A convergence zone has formed in Puget Sound. Winds at Pt. No Pt. are north 10 knots but at Alki Pt. are south at 30 knots.

Strong wind gusts ahead of the thunderhead may be gale force or higher, and, not too infrequently, hail or snow pellets fall from the storm.

Fortunately, the Puget Sound region has few thunderstorms, and most do not approach the ferocity of storms found east of the Cascades or in the U.S. Midwest. About six thunderstorm days a year occur in western Washington, but this may vary from none to over 40 thunderstorm days.

Thunderstorms in western Washington are mostly connected with strong cold or occluded fronts during winter and spring. However, on rare occasion, moist and unstable air from California may reach this area, bringing very active thunderstorms as it passes over western Oregon and Washington. Although the odds of getting hit by lightning are small, we should try to avoid operating under a thunderstorm if at all possible in an open boat.

Fog

Fog is quite common to northwestern Washington and is most likely to form under light wind conditions and clear skies. Fog may form either in high pressure or low pressure weather systems as long as there is sufficient moisture in the air to be cooled to the point of condensation. However, the most persistent fogs exist in high pressure weather patterns because the winds are usually lighter, and skies are more likely to be clear at night, allowing the heat from the earth's surface to radiate into space without being partially absorbed by clouds. A fog layer is typically 200 to about 500 feet thick. Fogs occurring over waters of western Washington are of two types: sea fog and radiation fog.

Sea fog forms over the ocean when a layer of air is cooled from below by cold water. It is very common along the northern California coast during the summer and early fall months and also forms along the Oregon, Washington, and British Columbia coasts during this time. When a high pressure center is off our coast, the wind blows into the Strait of Juan de Fuca and will bring a fog bank with it if there is one offshore. Sea fog, once it has formed, can persist over water in winds over 25 knots. We can see this by examining the sea fog situation in Figure 4-6.

This fog is most prevalent in the Strait of Juan de Fuca, but can reach southern Puget Sound on occasion, though it is more likely that this area will have a low cloud layer of stratus instead. Seas may be very rough in the eastern end of the Strait off Race Rocks and Smith Island, especially from late afternoon to around midnight when the sea breeze is strongest during the summer months.

If you are planning a trip from southern Puget Sound into the Strait of Juan de Fuca, you can be in 70°F weather under clear skies and a light breeze in Puget Sound, but in 50°F foggy weather with zero visibility and 25 to 30 knots of westerly wind when you get to the eastern end of the Strait.

Radiation fog is the prevalent type of fog in the Inland Waters. It derives its name from the fact that heat is radiated directly into space from the ground at night if there is no cloud layer to help keep the heat in. If the ground cools enough, water vapor condenses into droplets in the layer of air near the ground. Typically, radiation fogs do not occur very often in summer, but gradually increase in frequency toward the end of summer, reaching a peak around October.

Figure 4-7 shows the frequency of heavy fog occurrences at places where this type of weather record has been kept, and Figure 4-8 shows an actual day in a

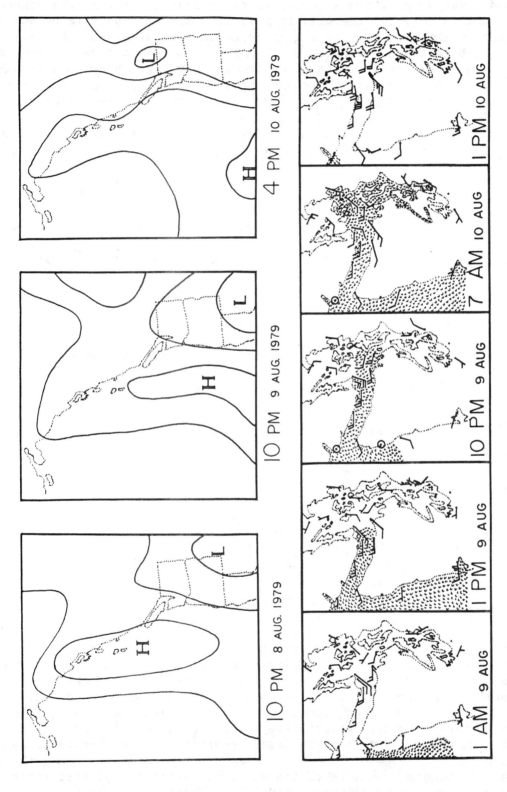

Figure 4-6 Sea fog episode in the Strait of Juan de Fuca and Inland Waters. All times PST. The top three panels show location of highs and lows. A high off the coast with lower pressure inland will cause westerly winds in the Strait of Juan de Fuca. The bottom panels show the winds and actual location of sea fog (stipled area), which blew into the Strait of Juan de Fuca from the ocean. Westerly winds of 20 to 25 knots occurred in the eastern end of the Strait, making seas quite rough. The fog finally dissipated by 1:00 p.m. on Aug. 10.

radiation fog episode that lasted over a week in the Inland Waters.

Radiation fog, in the late summer and early fall months, usually forms late at night and dissipates within a few hours after sunrise; but as the nights grow longer and colder, the fog will form earlier and dissipate later and later until it lasts the whole day. If fog is going to form at night, it will usually occur about two hours after dew has formed on exposed metal objects. Radiation fog drifts over the water from the land on light winds and is likely to persist longer over the water because the land heats up much more during the day, which helps dissipate the fog. Figure 4-9 compares a sea fog episode with a radiation fog episode.

The primary risks in fog are the accidental grounding of your vessel because you are lost or disoriented and the possibility of collision with another vessel. Most small boats, unless they have radar reflectors, do not readily show up on ships' radars. Depending on their size and construction material, they may be detected at less than one mile's range or not at all if the radar is not precisely tuned. To a ship, a one mile target distance is very close.

The weather summary broadcasted on VHF Weather Radio along with the forecasts will give you an idea of whether or not fog is likely, but like the winds, it is not possible to forecast exactly where the fog will occur or how bad it will be.

Sea Breeze in the Strait of Juan de Fuca

During the summer months when storm systems are weak and high pressure weather patterns settle in for long periods of time, the land heats up considerably,

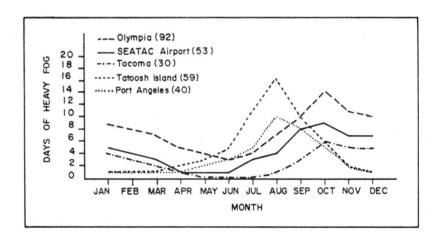

Figure 4-7 Frequency of occurrence of heavy fog in northwestern Washington. Fog is heavy when the visibility is 1/4 mile or less. The graph is based on years of records from stations that regularly recorded the occurrence of heavy fog. The numbers in parentheses in the legend are the average number of days in which heavy fog was observed either part or all of the day for the year (e.g., Olympia has 92 days in which heavy fog was observed at one time or another during a year). Note that the peak occurrence in the Strait of Juan de Fuca (Tatoosh and Port Angeles) is in mid-summer and at stations along the Inland Waters it occurs in October. Sea fog prevails in the Strait, and radiation fog is the dominant type in the Inland Waters.

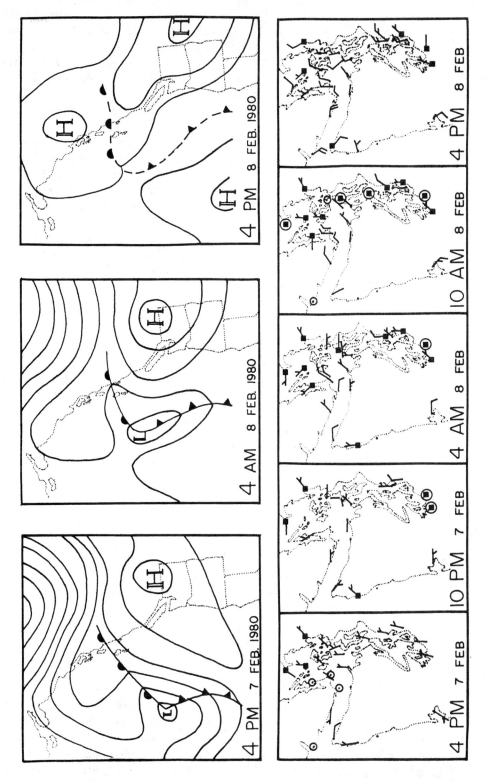

Figure 4-8 Radiation fog episode in the Inland Waters. All times PST. The top three panels show locations of highs and lows. A high dominates the weather pattern as the low off the coast disappears along with the front by 4 p.m. on 8 February. The bottom five panels show the stations reporting fog (black squares) over a 24-hour period. Wind speeds are shown using standard symbols explained elsewhere in this book. A circle around a box denotes a calm wind. This particular foggy period lasted over seven days before a front finally came through and "cleared the air."

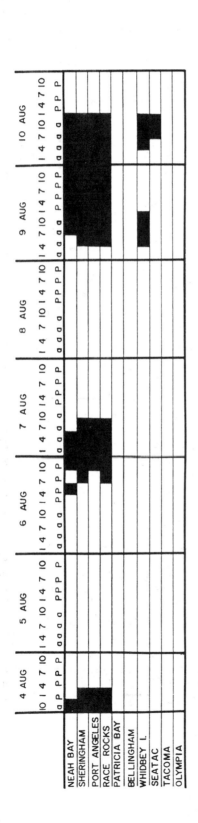

Figure 4-9 Sea fog episode compared to radiation fog episode. Times are PST where "p" denotes p.m. and "a" denotes a.m. A shaded block indicates fog was observed at the station. Top panel: Sea fog observed by stations in the Strait of Juan de Fuca over a 6 day period in August 1979. Note that with the exception of a few hours on 10 August, no sea fog was observed in the Inland Waters.

Bottom panel: Radiation fog observed over an 8 day period during February 1980. Note that the vast majority of the fog was observed at stations in the Inland Waters.

especially in eastern Washington. This heated air rises and is replaced in western Washington by air from the ocean by westerly winds which increase late in the afternoon and on into the night. If a high pressure system lies off our coast, the effect of this sea breeze is to increase the speed of the prevailing wind set up by the pressure system itself. At times, winds over 25 knots with gusts up to around 40 knots occur in the region east of Port Angeles. Fog and low clouds with chilly temperatures may come in with the sea breeze. Winds vary from northwest to southwest throughout the Strait and are likely to be stronger the closer you get to the Canadian shore. We can get a feel for the winds associated with the sea breeze by looking at an actual case (Figures 4-10 and 4-11). The sea breeze picks up gradually during the day, becoming strongest from about 7:00 p.m. PST to about midnight.

How strong it actually will be depends on the location of the high pressure systems and the pressure gradient through the Strait. If the high is directly over western Washington instead of offshore, the sea breeze will not be as strong since the resulting wind is not produced by a large push from the ocean along with the additional onshore flow caused by heating of the land. If the high center lies east of us, say over the Cascades or eastern Washington, the pressure will cause the natural wind to blow from east to west and will only be partially counteracted by the onshore sea breeze. Winds are often very gusty with sea breezes. These sea breeze winds may pose unacceptable risks to you during the months of July, August, and early September in the eastern part of the Strait of Juan de Fuca.

The Ediz Hook Eddy

In the region bounded by Ediz Hook (Port Angeles), New Dungeness Spit, Smith Island, Discovery Island, B.C., Trial Island, B.C., and Race Rocks, B.C., we often find westerly winds at Ediz Hook, southerly winds at Smith Island, and northerly winds at Discovery, Trial, and Race Rocks on the Canadian side of the Strait. We will refer to this counterclockwise circulation of the wind as the Ediz Hook eddy because there is no official name for this phenomenon.

At times, winds at both Ediz and Dungeness are 5 kn. or less from different directions; at other times, they may be 5 kn. at Ediz and over 25 kn. at Dungeness, which may cause quite a surprise as your traverse from smooth seas to very rough ones in a short distance. Anytime a weather front or low approaches the Washington coast, an increase in the easterly winds in the Strait can be expected along with an increase in the winds in the Ediz Hook eddy. Figure 4-12 shows several actual cases of the eddy so that you can see how it varies. Winds on the U.S. side may be just the opposite of those on the Canadian side of the Strait. The eddy is not likely to form when the wind is from the west in the Strait of Juan de Fuca, such as when there is a high pressure system just off the coast causing an onshore wind to blow.

When the air is flowing out the Strait, however, the eddy forms due to nearby hills and mountains disturbing the stream of air as it tries to turn the corners at Pt. Wilson and the southeastern tip of Vancouver Island. The barometric pressure at Bellingham must be higher than at Quillayute for this to occur. The eddy forms regardless of whether or not the winds are northerly or southerly in the Inland Waters south of Admiralty Inlet, but is strongest when winds are southerly.

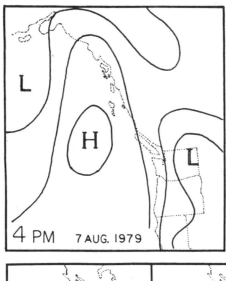

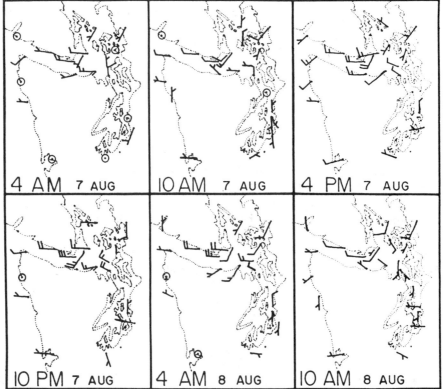

Figure 4-10   Sea breeze in the Strait of Juan de Fuca. All times in PST.
Top Panel: The strongest sea breeze occurs when a stationary or slow moving high pressure center is off the Washington coast, and a low is over eastern Washington. The onshore flow of wind from the high is increased by the sea breeze wind.
Bottom Panels: Wind speeds and directions during a typical sea breeze day. Fog and low clouds may be blown in from the ocean. Westerly winds may become quite strong, such as 30 knots for Race Rocks at 10:00 p.m., 7 August. Western Washington may experience sea breeze situations lasting more than a week at a time.

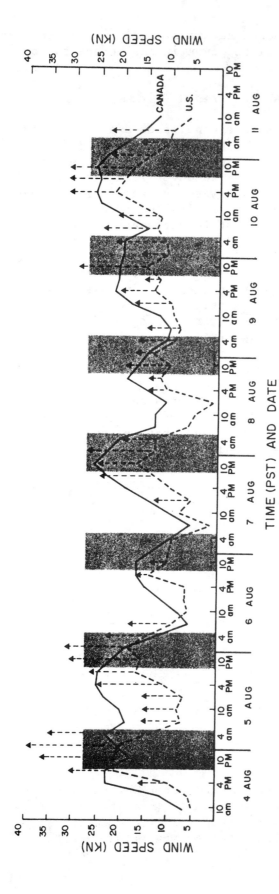

Figure 4-11 Sea breeze in the Strait of Juan de Fuca. All times in PST. Graph of the average sustained sea breeze winds on the Canadian side of the Strait of Juan de Fuca (Sheringham Pt., Race Rocks, Trial I.) and the U.S. side (New Dungeness, Smith I., Pt. Wilson). Solid line is for Canadian side; dashed line for U.S. side. Vertical shaded area shows nighttime hours; vertical arrows are measured maximum wind gusts from U.S. stations. (Canadian stations did not measure wind gusts). Note that average winds are stronger on Canadian side and that the sea breeze increases in the afternoon, reaching a maximum shortly before or after sunset. A 39-knot gust occurred at 1:00 a.m. 5 August.

Because of the general nature of marine forecasts, no specific mention of the Ediz Hook Eddy will be given in the forecasts for Strait of Juan de Fuca.

Hazardous East and Northeast Winds

On occasion, hazardous wave conditions are created in the Strait of Juan de Fuca and Inland Waters north of Whidbey Island when a high pressure system settles in over British Columbia and a low develops or moves into any area off the northern California to Vancouver Island coasts. During the late fall to early spring months, this may set up an arctic air outbreak (see Figures 1-9 and 1-10). At any rate, there is usually some colder air that flows into western Washington, and east to northeast winds 25 knots and higher may persist for days due to the large difference in pressure between Bellingham and Quillayute. The slower the high and low centers move, the longer these winds will persist. The east to northeast winds shown in Figure 4-13 lasted from 13 to 19 February 1980 although the colder air in this very mild arctic air outbreak only lasted the 14th and 15th.

Temperatures in northwestern Washington dropped from the low to middle 40s °F to the upper 20s to lower 30s °F as the cold air from Canada flowed over the area. The wind chill temperature hovered around -2°F in the areas of 30 knot winds (see Wind Chill Chart in Appendix). Waves of around 6 feet persisted in the northern Inland Waters and western end of the Strait of Juan de Fuca. Through the use of weather summaries on VHF Weather Radio or a couple of consecutive weather maps found in newspapers, we could have reasoned that the pattern was slow in changing and that we could have expected the winds to persist. Marine forecasts would have given us an idea of the strength of winds.

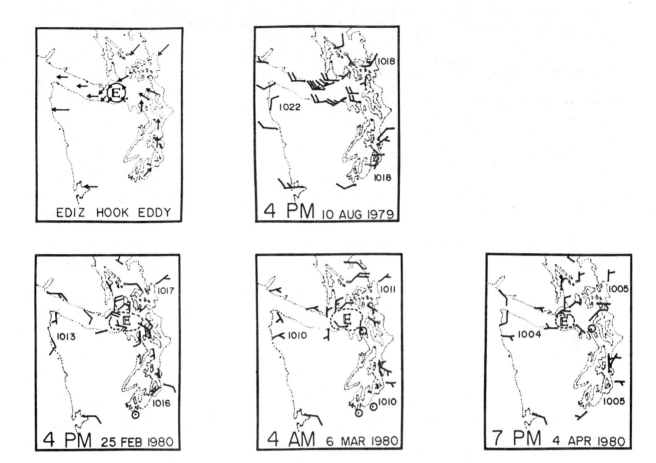

Figure 4-12   The Ediz Hook eddy. Top Left: The Ediz Hook eddy is a counter-clockwise circulation of the wind in the eastern end of the Strait of Juan de Fuca. It is very likely to occur when the barometric pressure at Bellingham is higher than at Quillayute and is unlikely to form when the pressure at Quillayute is higher than at Bellingham. The size of the eddy varies as can be seen in the actual cases shown above. It can occur anytime of the day or night during any month of the year if the conditions are right and can persist for over 12 hours.

Top Right: No Ediz Hook eddy. Note that pressure at Quillayute is higher than at Bellingham, which causes a westerly wind in the Strait of Juan de Fuca. An eddy is very unlikely to occur in this situation.

Bottom Left: Strong Ediz Hook eddy (shown by dotted line and "E"). Note that Bellingham's 1017 mb pressure is much higher than 1013 mb at Quillayute. Wind at Smith I. is southeast at 30 knots. Sometimes winds at Dungeness Spit will be over 25 kn. and less than 5 kn. at Port Angeles. All times are PST.

Bottom Middle: Weak Ediz Hook eddy covering entire eastern end of the Strait. Note that pressure at Bellingham is only 1 mb higher than at Quillayute.

Bottom Right: Weak Ediz Hook eddy covering a small area. Again the pressure at Bellingham is only 1 mb higher than at Quillayute.

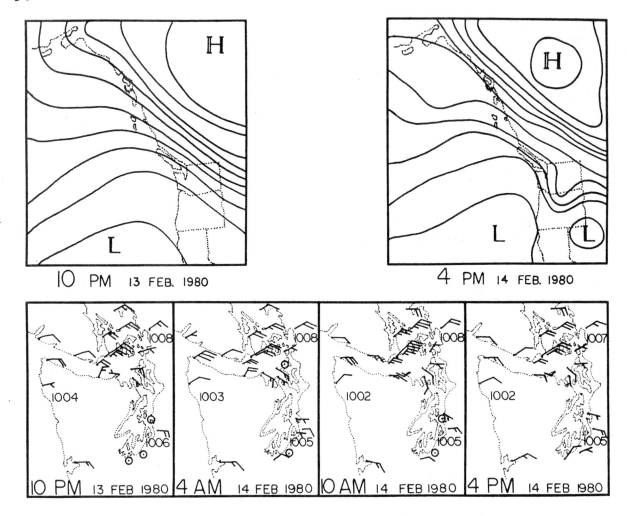

Figure 4-13 Hazardous east and northeast winds. All times PST. A high pressure system developed over British Columbia while a low developed off the northern California coast (top panels). This resulted in a large difference in barometric pressure between Bellingham and Quillayute as can be seen by the closely spaced isobars in the top panels and the actual pressure in millibars shown in the bottom panels. (Pressures at Quillayute, Bellingham, and SEATAC are shown.) A pressure difference of more than 3 millibars is likely to create winds over 25 knots in the northern Inland Waters and Strait of Juan de Fuca. Note that Bellingham's pressure was 6 mb higher than Quillayute's at 10 a.m. on the 14th. The wind at Neah Bay at this time was east 30 knots with gusts to 40 knots. Also note that an Ediz Hook eddy (see Figure 4-12) developed.

Chapter 5
COASTAL WEATHER

Like the Inland Waters and Strait of Juan de Fuca, the coastal area and waters offshore are affected by numerous weather patterns, only some of which pose risks to the mariner. This chapter will be limited to those weather patterns presenting high risk to small craft, but in some cases, to very large craft also. These situations almost always involve high winds and seas due to tight pressure gradients. In other words, the isobars are close together, which is just another way of saying that a large change in barometric pressure occurs over a given distance. Figure 5-1 shows what is meant by pressure gradient over the open ocean.

The forecaster attempts to predict where and when these tight pressure gradients will occur but is sometimes frustrated in his efforts because of the small amount of weather data available from the open ocean. Areas the size of Alaska may have no observations at all in the eastern Pacific Ocean. Except for an occasional ship, tugboat, or fishing vessel report, there are seldom any weather observations at all over the ocean region between the coast of Washington and an automatic environmental data buoy anchored about 250 nm offshore at (46.2°N, 131.0°W). There are six weather reporting sites on the coast between Cape Flattery and Cape Disappointment (if all the equipment is working). Satellite pictures of the cloud patterns help quite a bit, but the lack of surface observations over the water makes it very difficult, if not impossible, to forecast for small local areas over the open ocean with the same degree of confidence done over small areas of land. Generally, marine forecasts are likely to be less accurate than those made for land areas.

Weather fronts, lows, and highs usually, but not always, come from the SW-W-NW sector and move E to NE. In most cases, the ocean (and land) is affected by a moving series of weather systems, but sometimes lows or highs may stall over the coast and we get long periods of "bad" weather or "good" weather. We will concern ourselves here with the "bad" weather situations, pointing out the magnitude of wind and wave forces associated with various situations, and we will also look at the problem of fog along the coast.

Strong Winds with Lows

Low pressure systems, which can bring devastating winds to the coast and interior of Washington, may travel over the ocean at speeds from about 5 to over 60 knots. Their speeds are not constant and change with time. Movement of both lows and highs is slowest during the summer months and much faster the

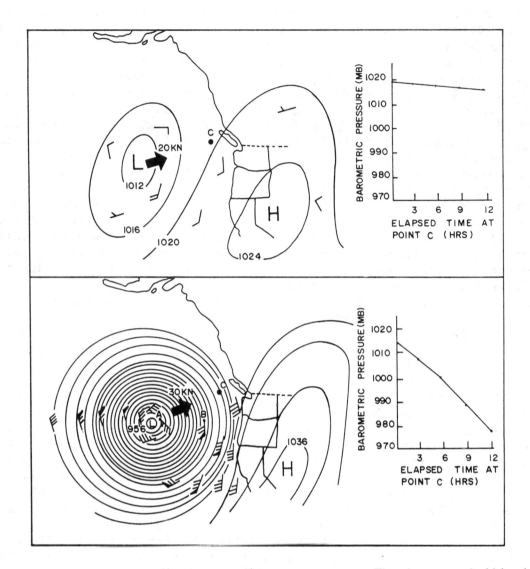

Figure 5-1 Pressure gradient over the open ocean. The top panel illustrates a weather pattern in which a weak low (about 1010 mb) is adjacent to a high pressure system of about 1026 mb. The isobars are not close together either around the low or high, so the pressure gradients are weak. Winds are 20 knots or less. In the bottom panel, an extremely deep low (956 mb) is adjacent to a strong high (about 1038 mb). The pressure difference between the two systems is 82 mb, and a large pressure gradient exists, which is shown by the closely spaced isobars. Winds in this case are gale and storm force and will be strongest in the quadrant between the high and low.

In a low, winds may very well be strongest several hundred miles away from the center unless the pressure gradient is extremely high closer in. Refer to the bottom panel in which point A is very close to the low center and point B is much further away where the curvature of the isobars is not nearly as small as at point A. In a low, for a given pressure gradient (e.g., equal spacing of isobars), the winds will be weaker nearer to the center than they are farther out. In fact, they may be twice as fast several hundred miles away as they are

other months of the year. This is due to the jet stream. To understand why storms behave as they do, it is necessary to know about the jet stream.

The jet stream affecting our area is just one of several major jet streams of the Northern Hemisphere. It is a fast flowing river of air separating warm air masses to the south from the cold ones to the north. It is often more than 5,000 ft. thick and hundreds of miles wide, being found between about 18,000 ft. and 40,000 ft. Highest wind speeds are usually found at around 30,000 to 35,000 ft. and are greatest during the fall through spring months in which speeds to around 250 knots have been observed. This stream of air meanders in a wavy manner around the Northern Hemisphere in a general west to east flow. (It should be pointed out that the Southern Hemisphere also has jet streams.) However, the direction is not constant, and when the jet reaches from the polar regions into the heartland of the U.S. during winter, extremely low temperatures occur. Similarly, when the jet stream extends from the tropics into Alaska or other northern latitudes unusually high temperatures result. Sometimes the jet stream splits into two streams. Where this occurs, weather fronts will be torn apart, making for generally fine weather under the "split." Figure 5-2 shows an example of a jet stream.

The jet stream plays a very important role in storm development. Unfortunately, without radiofacsimile recorders it is not possible to get information on the jet stream aboard vessels at sea. We can, however, sometimes use weather signs to detect its presence.

This is done by noticing changes in high and middle clouds. If you can detect definite movement in cirrus clouds, the jet stream at cloud level is about 90 knots or higher. For elevations below about 20,000 ft., this indicates a strong jet stream. It will be much stronger at 30,000 ft. If the clouds come from a southerly or westerly direction and grow thicker with time, followed by altocumulus and altostratus clouds, watch out! A storm may be arriving in 12 to 24 hours. Be sure to watch for falling readings on your barometer and tune into the weather broadcasts. If the forecast does not go along with what you are observing, then believe Mother Nature's warning and be prepared for a blow. Clouds coming from a northerly direction do not usually signify an approaching storm.

A low that crosses underneath a strong jet stream will deepen, causing the winds around the low to increase. The jet stream helps determine which way the

⬅━━━━━━━━━━━━━━━━━━━━━━━━━━━━━━━━━━━━━

Figure 5-1 caption continued

within 60 or 100 nm of the low center. Suffice it to say that the centrifugal force on the flow of air is greater near the center than it is farther away. This force partially counteracts the effects the air pressure difference has in causing the wind and reduces the wind speed. Just the opposite effect occurs in a high, but because the pressure gradients are much weaker, the increase in wind speed due to curvature of the isobars is usually not too important.

Finally, refer to point C in both panels. As the low moves toward this point, the pressure readings on a barometer change much slower with a weak low than with a deep one. The 12 hr. change in pressure is shown in both cases.
Rapidly falling (or rising) pressures as shown in the bottom panel are usually indicative of dramatic changes in the wind and weather.

low moves and how fast, but the low does not follow underneath the jet stream like a leaf in a creek. The path actually followed by a low and weather fronts is a lengthy topic in itself. Suffice it to say that lows and fronts will move in the same general direction of the jet stream. We would not find, for example, a low moving SE in a jet stream flowing toward the NE. The low would

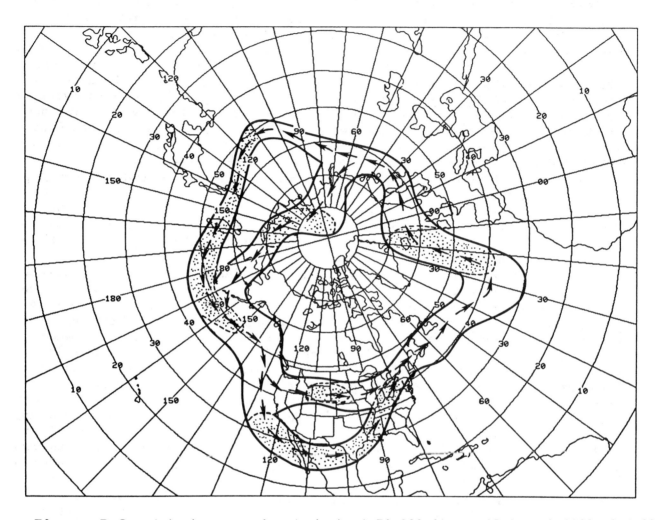

Figure 5-2   Jet stream as found at about 30,000 ft. on 12 August 1982 at 4:00 p.m. PST. Stipled areas denote areas of 70 to 90 knots, and the other areas are 30 to 69 knots. Because it is summer and there is not the great difference in temperature between air masses of the north and south, the jet stream is fairly weak. A winter time example would show areas of winds well over 120 knots. Notice how the jet generally flows west to east, but that it occasionally dips southward or curves northward. A ridge of high pressure exists over the Gulf of Alaska in the figure, while low pressure exists over British Columbia. Inflections in the jet alternate from trough to ridge to trough all around the Northern Hemisphere. Of course, the pattern shown in the figure does not stay constant and will undergo remarkable change with time. Where there was once a trough, a ridge will move in, and vice versa. Surface low pressure systems deepen in the trough areas, while highs strengthen under influence of upper level ridges.

mostly follow a northeasterly course but may spin away from beneath the jet stream and follow a more northerly trackline. Often times, particularly in the Gulf of Alaska, storms spin away from the jet stream to travel north, then drift around in a counterclockwise loop toward the west then toward the south and finally back toward the east. When a surface low decouples from a jet stream, it weakens with time.

Super-storms

The strongest lows always come from a southerly direction and have their origins in the subtropics or tropics, sometimes more than 2,000 miles away. Occasionally, during the fall through spring months, some of these storms develop into what we will call super-storms. Most of these race into the Gulf of Alaska where they eventually die, but on occasion they move NE to N along the Oregon/Washington coast, playing havoc with inhabitants living west of the Cascades and sometimes much farther inland. Super-storms develop extremely rapidly with central pressures dropping down below 970 mb (some go below 940 mb). Even if the storms do not hit the coast with storm force winds, high swells generated by them can still cause considerable problems, especially during high tide. For example, a 936 mb low in the northeast Gulf of Alaska brought 35 ft. waves to the Oregon coast on 29 October 1977. One woman was swept to sea from her motel room. A low this deep in the Gulf of Alaska will usually cause problems along the Pacific Northwest coast.

Some recent examples of super-storms are the Columbus Day Storm (12 October 1962), the Hood Canal Storm (13 February 1979), and the Friday the 13th Storm (13-14 November 1981). All of these storms had their origins in the subtropical or tropical regions of the Pacific and developed into very intense lows once they moved northward and mixed with the air on the cold side of the jet stream. We will go into some detail on these storms because they are truly shipwreckers and something any mariner would prefer to miss. Careful study of these examples will show the danger they pose. The storms deepen at a phenomenal rate, and because of this very rapid development, mariners may have less than a 12 hour warning to take evasive action. Wind speeds over 100 knots with higher gusts are possible in such storms. Figures 5-3, 5-4, and 5-5 illustrate four super-storm events -- the three mentioned above and one in the Gulf of Alaska.

Fortunately, such storms are extremely unlikely during the summer months, but once October comes, the risk increases considerably that such a storm (or series of storms) will occur somewhere in the northeastern Pacific Ocean. The area of ocean covered by gale and storm force winds may vary considerably. Figure 5-6 compares the high wind area of the Gulf of Alaska storm with that of the Friday the 13th storm.

Although each storm has to be evaluated on its own peculiarities, there are similarities between storms so that studying an example in detail will give an idea of the magnitude of forces involved. Figure 5-7 shows some details about the Friday the 13th storm. It is obvious that where we are in relation to the storm determines not only wind direction and speed, along with wave heights, but also duration of the storm in respect to our location. If we had been west of 140°W in this case, the effects of the storm would not have been significant because the low tracked NE away from us. On the other hand, if the low center

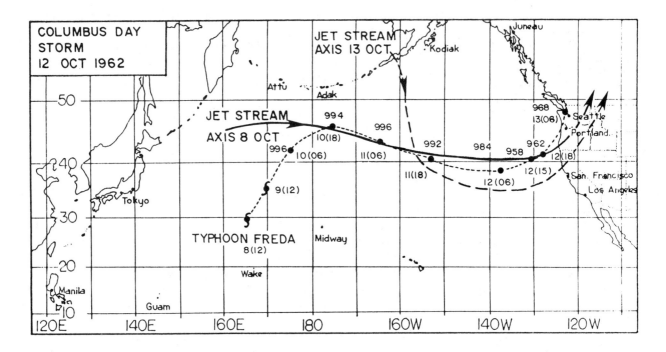

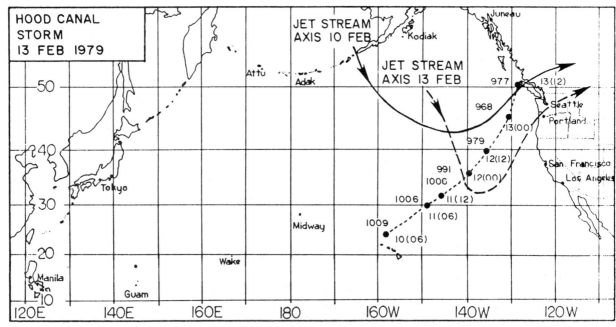

Figure 5-3 Tracklines of the Columbus Day Storm and Hood Canal Storm. Pressures in mb are shown for each position of the low center (dots denote position, not size) and the date and time (GMT) are shown by each position. The main axis of the jet stream is shown at or near the start of the period when the low center was detected and at or near the time that the low center reached its lowest pressure. Note that the low deepens very fast once it passes underneath the jet axis and crosses into the colder air on the other side. The Columbus Day Storm originated as a typhoon, but was not one when it got north of about 35°N. The Hood Canal Storm originated as a very weak

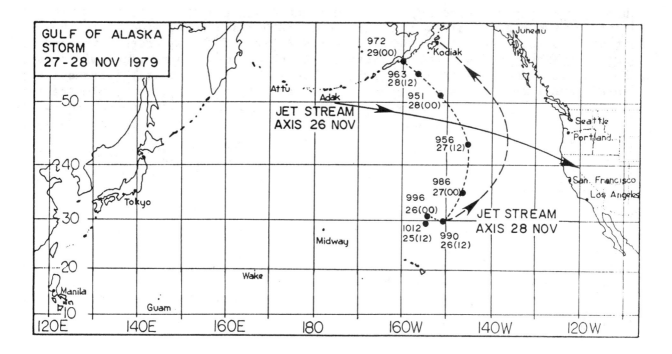

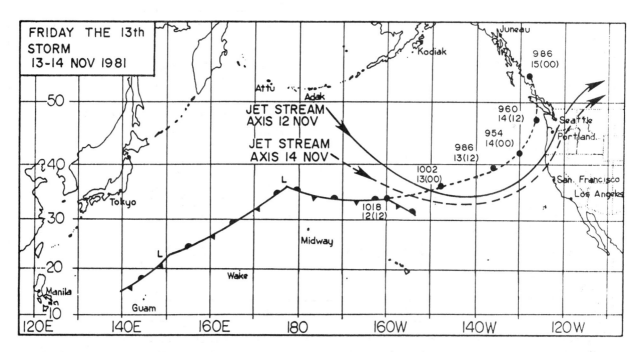

Figure 5-4 Tracklines of the Friday the 13th Storm and a Gulf of Alaska super-storm. Note the great change in direction of the jet stream during development of the Gulf of Alaska storm.

Figure 5-3 caption continued:
1009 mb low over Hawaii. It should be mentioned that lows which we would call weak at our latitude may not be so in the tropics because the air is much moister and winds blow harder with a given pressure gradient.

had tracked right toward us, we not only would get the strong NE to SE winds ahead of the center (along with very poor weather and gusty winds with the fronts) but also the strong SW to NW winds after the center passes overhead. The wind around a low is not constant, and pockets of much stronger winds may be experienced in the prevailing wind. These pockets of greatly increased wind speed cannot be forecasted because of the lack of data. Inside the low center itself, which may measure only about 40 to 60 nm across in a super-storm, winds may be variable in direction and only blow at 10 to 20 knots. Figure 5-7 shows the super-storm of 13-14 November 1981 in detail.

Weather Fronts

Weather fronts associated with lows deeper than about 980 mb will bring very inclement weather and high winds. However, even "weak" fronts will bring an increase in wind and cloudiness at least, and so when we hear that there is a front off the coast, it is wise to consider the following: (1) How low is the pressure in the low pressure system in which the front is imbedded? Words such as "intense," "deep," "strong," etc., used to describe the low and front in the

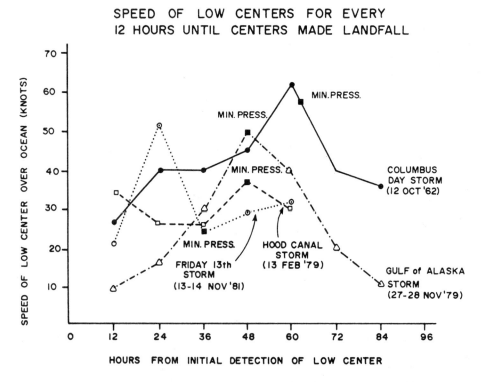

Figure 5-5 Some facts about four super-storms. Note that during the 12 to 24 period before a low reaches its lowest pressure, the speed of the low increases dramatically. The speed of advance then usually slows down after the low reaches lowest pressure. For example, in the 24 hours before minimum pressure was reached on the Friday the 13th Storm, the speed of the low center over the ocean increased from 20 knots to 50 knots and then fell back to about 20 knots after minimum pressure was achieved.

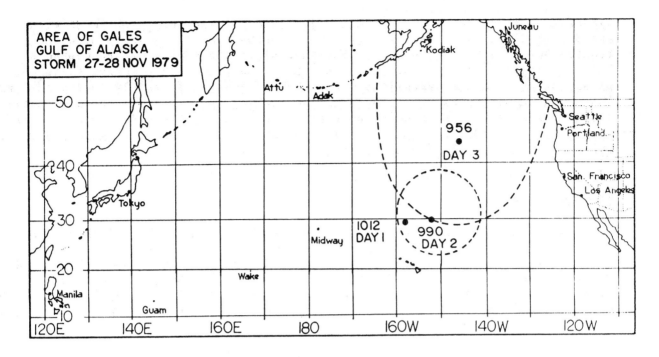

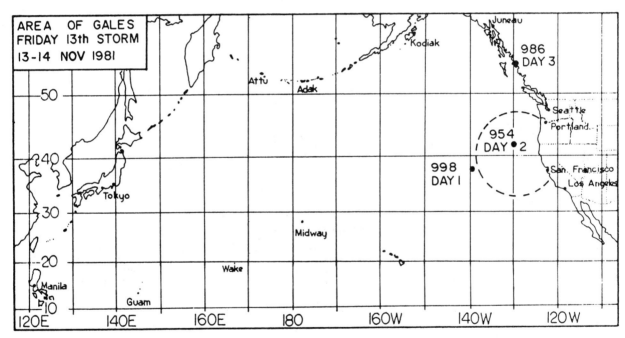

Figure 5-6 Ocean area covered by gale and/or storm force winds for two super-storms. Note that both storms had low centers with nearly equal pressures (956 mb Gulf of Alaska Storm and 954 mb Friday the 13th Storm). The strong wind area in the first case grew extremely fast within 24 hours and covered the entire Gulf of Alaska and waters off the West Coast. The ocean area covered by the Friday the 13th Storm did not grow nearly as much because the low moved near land and made landfall quickly. This weakened the low because of friction and loss of energy from moist ocean air.

BUOY OBSERVATIONS

4 P.M. PST 13 NOV 1981

Buoy #	Wind	Baro.	Waves Ht.	Per.
2	SE 18	967	18	9
6	NW 28	996	15	14
10	E 22	994	10	14
5	NE 20	987	15	14

Buoy #	Wind	Baro.	Waves Ht.	Per.
2	W 34	995	21	11
6	WSW 22	1001	15	14
10	SSE 48	973	26	14
5	NW 28	987	18	14

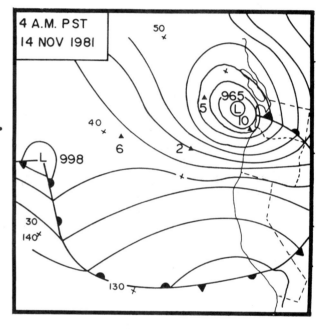

4 A.M. PST 14 NOV 1981

4 P.M. PST 14 NOV 1981

Buoy #	Wind	Baro.	Waves Ht.	Per.
2	WSW 10	1005	13	14
6	W 12	1000	12	14
10	SW 22	1007	23	14
5	SW 18	999	13	14

Figure 5-7 The super-storm of 13 and 14 November 1981. Three surface weather maps spaced 12 hours apart are shown with isobars drawn at 4 millibar intervals. The central pressure of highs and lows are shown near each center. Data buoys anchored off the Oregon and Washington coasts are shown as small triangles with the buoy number written nearby. (The complete buoy numbers are 46002, 46006, 46010, and 46005.)

4 PM, 13 NOV: The low has reached its lowest central pressure (954 mb), and the cold front has overtaken the warm front from about the northern California coast to near the low center. Lows reach their lowest pressures during this process and begin to weaken once the warm front has been completely overtaken. Note that the isobars are very close together, indicating gale to storm force winds.

4 AM, 14 NOV: Low has started to weaken, but is still very deep and packing storm force winds. Note that another low has started to develop about 1000 nm SW of the original low. This is very typical with original lows that have central pressures lower than about 980 mb. The deeper the original low, the stronger the secondary low will be. The secondary low usually follows a similar path to that of the first low, but the trajectory is usually displaced further eastward. In the case above, the secondary low moved E as shown, then curved NNE after 4:00 p.m. on the 14th and followed a track parallel to that of the original low but 60 nm further east.

4 PM, 14 NOV: The original low has moved onshore and weakened considerably. The secondary low continues to move east and shortly after 4:00 p.m. curved to the NNE and deepened to 985 mb. It struck the Washington coast on the 15th.

Note: Wind speed in knots; barometric pressure in millibars; significant wave height in feet; wave period in seconds. Wind speed measured by environmental buoys is averaged over an 8.5 min. period and will be lower than that reported by a person reading an anemometer because he is not likely to average the dial readings over this long period. Also, the anemometer on a buoy is often sheltered from the wind when the buoy is heeled over in a wave trough. Winds of 106 knots were recorded on the southern Oregon coast and around 50 knots with gusts over 60 knots on the Washington coast with this storm.

weather report should put us on guard; (2) How fast is the barometer falling? Rapid falls of 5 mb or more in 3 hours indicate a great change in the weather and winds; (3) What time of the year is it? Fronts are especially strong from late September until about June; (4) Cloud types? What types of clouds we observe gives some indication where we are in relation to the front (see Chapter 1).

Table 5-1 lists the typical effects associated with warm and cold fronts over the open ocean. The cold front typically takes 36 to 48 hours to completely engulf the warm front, during which time the low continues to deepen. Once this process is completed, the low will start to weaken as is illustrated in Figure 5-7. Thus, if we know what kind of front is approaching, we can get some idea about the stage of development of the low center also. Until the cold front has completely overtaken the warm front to form an occluded front, the low will continue to deepen.

Weather conditions around fronts follow a fairly predictable sequence over the ocean and are shown in idealized form in Figure 5-8. Comparison of this diagram with an actual case shown in Figure 5-9 shows that the idealized version is useful in indicating general weather conditions over the ocean around weather fronts. The big problem -- one which is often difficult -- is predicting the magnitude of the winds and waves along with precipitation and visibility.

When a weather front approaches the coast, southerly winds ahead of the front often get squeezed between the coastal mountains and the front itself, which causes an increase in the wind speed in this zone. Figure 5-10 shows an actual example of this occurring.

In some cases, the winds will be strongest before the front passes than afterwards. Without ship and buoy observations on weather maps, we have no way of knowing for certain whether the winds will be strongest before or after frontal passage. However, if we know that the front approaching us is associated with a deep low and the winds do not pick up to at least gale force before the front arrives, then we can very likely expect stronger winds after the front passes by. Conversely, if the winds are gale force or higher ahead of the front, then the winds behind the front are likely to decrease. Wind speed and direction before, during, and after frontal passage vary widely. In less than one minute, wind direction may vary much more than 90° and wind gusts may blow 50% higher than the average wind. Such conditions with strong fronts are disastrous to sailboaters.

Like lows, fronts are weakest during the summer months, but near the end of summer in early September, strong winds can occur in the nearshore area off the Washington coast in such a narrow band that they may escape detection by the regular network of weather stations. It is possible that no marine warnings will be in effect until after the event has started in such cases. One event of this type occurred over the 8th and 9th of September 1978.

Deceptively calm weather had persisted off the coast the preceding week, luring many small boats far to sea. A weak cold front associated with a 1000 mb low approached the coast on the morning of the 7th and was about 300 miles offshore when a disturbance in the upper atmosphere traveling at a greater speed overtook the surface low and front. Within 24 hours, the pressure in the surface low dropped 12 mb, setting up a tight pressure gradient in the region 5 to 10 miles off the beach to about 90 miles offshore. The resulting gales and waves in this narrow band sank 6 boats and claimed two lives. The forecasts did not predict this to happen, so always keep a wary eye on Nature's signals,

TABLE 5-1   FRONTAL WEATHER OVER THE OPEN OCEAN[a]

	Warm Front Approaching	Warm Front Passing	Warm Air Sector	Cold Front Approaching	After Cold Front Passes
BAROMETER READING	falling	stops falling, may slightly rise	little change, then falling	falling to rapidly falling	rising
WIND	SE to S, increasing	S to WSW with possible great increase	steady and strong; possibly becoming SE to S as cold front nears	strong SE to S	rapid change to SW to NW; may increase
TEMPERATURE	rising slowly	rising	little change	little change	sudden drop
PRECIPITATION	becoming steady; drizzle to heavy rain or snow	rain or drizzle	rain, drizzle, fog	heavy rain; possible thunderstorms	heavy rain, hail or snow showers; then intermittent showers
VISIBILITY	deteriorating	poor	poor	improving rapidly	good, except in showers
CLOUD TYPES	Cirrus Cirrostratus Altostratus Nimbostratus	Nimbostratus Stratus	Stratus Stratocumulus	towering Cumulus and Cumulonimbus	Cumulonimbus Altocumulus Cumulus

[a] The effects listed in the table are the ones ordinarily observed. Sometimes another cold front follows several hundred miles behind the first cold front, particularly in higher latitudes. An occluded front may act like a cold front or a modified warm front, depending on the strength of the front and the temperature of the air behind the front.

particularly fast moving high and middle clouds from the south-southwest-west directions. Figure 5-11 shows how narrow this band of strong winds was. In conclusion, as September approaches, become much more wary about any lows or fronts approaching the coast.

Strong Winds with Highs

Although high barometric pressure is usually associated with weak winds and fair weather, there are occasions when winds and seas in a high can be just as hazardous as in a low. The two situations where this occurs are: (1) after the passage of a low when high pressure rapidly builds over the coast; (2) when a very strong high pressure system is off the coast, and a low pressure system is situated over eastern Washington or northern Idaho.

A high pressure system approaching the coast will bring northerly winds from the coast to several hundred miles offshore. A rapidly rising barometer would be indicative of a high building up along the coast and would be indicated by rises of 2 to over 4 mb in three hours. Figure 5-12 shows an actual example of such an occurrence along the Washington and Oregon coasts.

Over the ocean, a high pressure center of 1030 to 1039 mb should be considered to be strong, and one having a pressure of 1040 mb or more is very strong -- the average pressure of a high over the eastern Pacific is 1025 millibars.

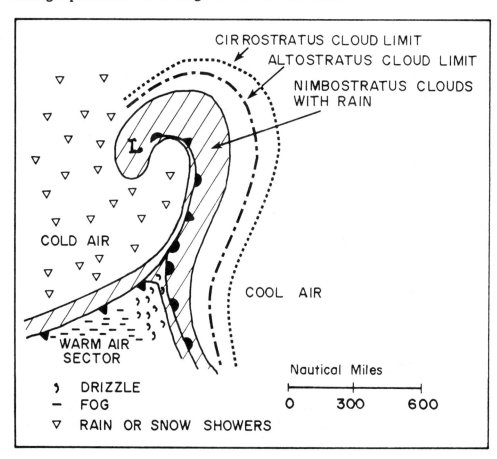

Figure 5-8 Weather and wind conditions around "ideal" weather fronts.

Consequently, if a strong high is within six or seven hundred miles of the
Washington coast, and a low is located east of the Cascades or in northern
Idaho, gale force northerly winds are likely from the coast to around 400 nm
offshore. If the high pressure center is more than six or seven hundred miles
offshore, winds are not likely to be gale force but in the 20 to 30 knot range.

It should be pointed out that this is a very common phenomena from the
southern Oregon coast to Pt. Conception near Santa Barbara, California -- par-

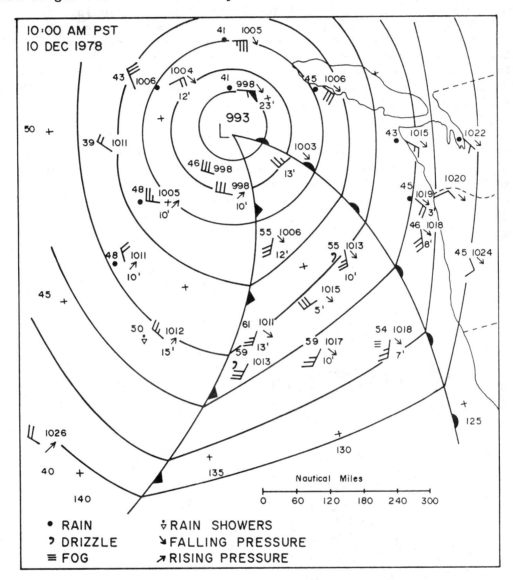

Figure 5-9 Observed conditions around a warm front, in the warm air sector,
and cold front at 10:00 a.m. PST on 10 Dec. 1978. The isobars are drawn every
4 mb. Significant wave height is labeled in ft., air temperature is in °F,
pressure is in millibars. Example using observation made just off the Columbia
River mouth: Wind SE 20 kn., barometer 1019 mb, air temperature 45°F, wind
waves 3 ft., rain, and pressure is falling.

ticularly from late spring to early fall. During this period, the high pressure system normally found off the California coast increases in strength and extends further northward. At the same time, low pressure forms in the Sacramento Valley and other areas of southern California due to heating by the sun. The result is often a tight pressure gradient between the high and low along the U.S. West Coast. Winds to over 50 knots and waves over 25 ft. can result from this, yet the skies are clear or only partly cloudy. This condition can persist for several days with no letup in the wind. A similar situation can develop off the Washington coast in which low pressure forms somewhere east of the Cascades. Figure 5-13 shows an actual case of an event in which this occurred.

Fog Over the Ocean

In Chapter 4, the conditions in which fog forms were discussed, and it was shown that the coast of Washington has more foggy days during summer months than other times of the year. Fog that occurs with a high pressure system is likely to last much longer than fog occurring with a low pressure system. The reason being that often times the high remains nearly stationary over an area with conditions favorable to fog formation, while the fog associated with a low usually occurs in conjunction with a moving weather front. Fog develops over the ocean due to a number of factors: the cooling of the air by the water (particularly in summer when prevailing northerly winds have the effect of causing water to move offshore to be replaced by colder water from the depths);

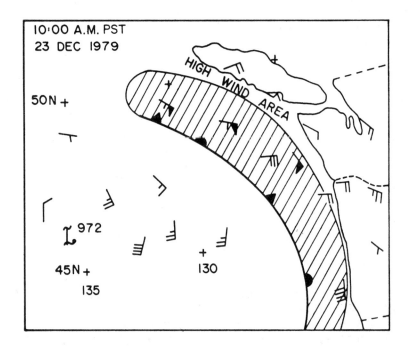

Figure 5-10   High wind area between an occluded front and the coast at 10:00 a.m. PST on 23 December 1979. Note that winds are up to 55 knots ahead of the front, which was moving northeastward, and only around 25 to 40 knots behind it. The 972 mb low center is a deep low and the front is a strong front.

the loss of heat from the moist air layer next to the ocean, especially on clear nights; addition of moisture to the air layer by rain or drizzle (important with weather fronts); light winds (usually between 2 and 20 knots, although fog can persist in winds over 30 knots). One other factor also contributes to the formation of sea fog and that is the trapping of this moist air in a layer, say 500 to 1,000 feet thick, due to a temperature inversion in the atmosphere. Normally, air temperature decreases with height, but on occasion it increases. The region of air in which this occurs is called a temperature inversion, which means that warmer air overlies cooler air, preventing the mixing of air from below with that above. In a high pressure system, air is sinking down from aloft, and through the effects of compression (like when you use a tire pump) the air is heated, resulting in a layer of heated air aloft. If this temperature inversion sinks close enough to the sea, it very effectively traps moisture in the lower atmosphere and raises the humidity as a result. Combine this effect with some of those mentioned above, and a very dense and persistent sea fog develops.

Sea fog may last for days and days during which visibilities will often be under a hundred yards. Sometimes, sea fog originates over land or bays and blows out to sea on easterly (offshore) winds. Unless there is already a sea fog along the coast, these land fogs may only extend 2 or 3 miles offshore where they may be blown southward as a "plume" of fog for 10 to 20 miles by the prevailing northerly winds just offshore. Easterly winds out of bays and inlets occur when the barometric pressure is higher east of the Olympic Mountains that it is along the coast.

Sea fog can also form hundreds of miles to sea and come rolling toward the coast on prevailing winds. Atmospheric conditions and water temperatures vary from place to place, so fog may exist over thousands of square miles or in small patches only a few miles across. It is difficult to accurately predict

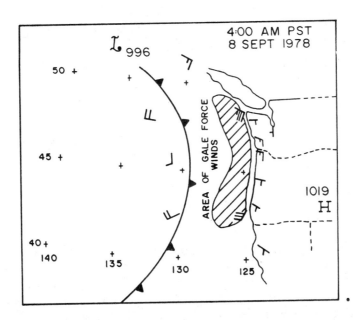

Figure 5-11   Narrow band of high winds near the coast ahead of a cold front at 4:00 a.m. PST on 8 September 1978. The front is moving eastward. Note that the winds on the coast are only 5 to 15 knots.

where sea fog will actually occur. About all we can do is understand what conditions cause it, and once it has formed, estimate how long it might last.

If a fog bank is observed offshore early in the day during the summer months, there is a good chance it will come ashore by afternoon because of the sea breeze effect (see Figure 5-14).

A fog bank at sea appears first as a thin white line on the horizon. As it draws nearer, it becomes a wall of gray and white cloud on the water. Sometimes the water is quite choppy in the fog if the wind bringing it in is more than 10 to 15 knots.

Because of the lack of weather observations over the open ocean along the

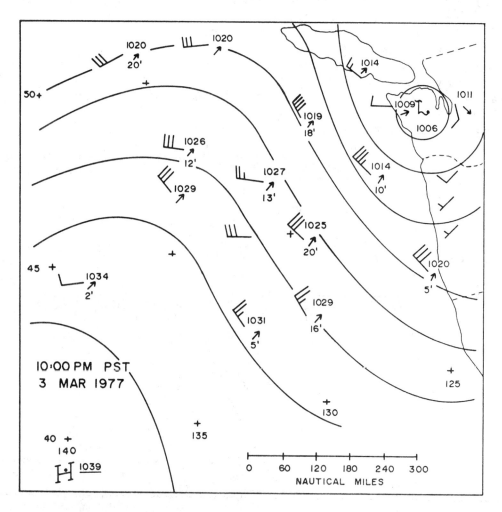

Figure 5-12   Strong northerly winds occur with a high pressure system building over the coast after a low has passed by. The wind speed on land may be deceptive as can be seen at Quillayute (WNW 10 kn.) compared to off the Washington coast (NW 40 kn. and significant wave height of 18 ft.). The arrows beside the wind barbs indicate the trend in the barometric readings. An arrow slanting upward indicates rising pressure and a downward arrow indicates falling pressure. Note that all the pressures are rising over the ocean and that winds are near gale to gale force to almost 400 nm offshore.

Washington coast, not much is known about fog conditions over the sea. What little data there are seem to indicate that sea fog is most prevalent with northerly winds (NW-N-NE) during May through September. Easterly winds (NE-E-SE) occur with fog mostly in the dead of winter, January to February. The fog associated with these winds blows from land to sea. Easterly winds, by the way, except during winter, are the least likely to be associated with fog since they blow offshore. Southerly winds (SE-S-SW) are the winds that occur with fog associated with fronts, being most likely August through March or April. Finally, westerly winds (SW-W-NW) with fog are most likely in the sum-

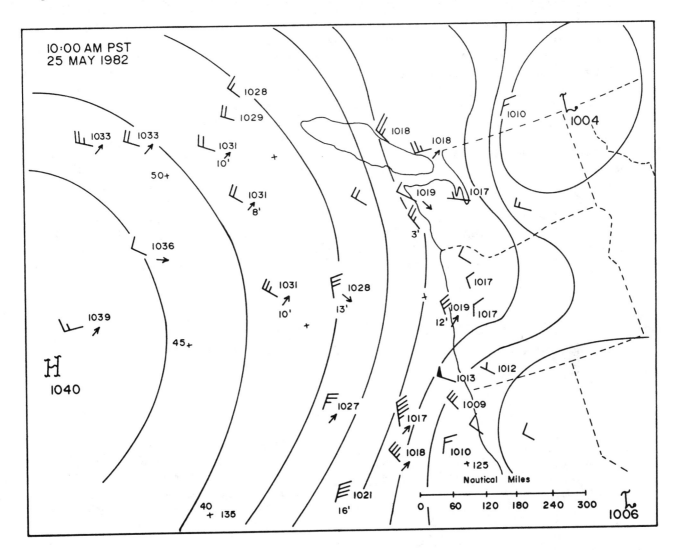

Figure 5-13 Strong winds off the West Coast due to low pressure east of the Cascades and a very strong high pressure center offshore. Note that the winds are particularly strong near the Oregon/California border. Along the Washington coast, winds are 20 to 25 knots. The barometer reading is written by most of the wind barbs. Although the upward slanting arrows indicate rising pressure over the ocean, in this case, the changes are very slight, say less than 1 mb per three hours. This is unlike the situation where high pressure rapidly builds up behind a low that has moved ashore.

mer months and again during the winter.

Fog can move up and down the coast in response to small changes in wind direction caused by very local weather conditions. To show how complicated this can become, a series of weather satellite pictures in Figures 5-15 and 5-16 show an actual sequence of events during a sea fog episode.

Over the open ocean, sea fog may persist from a day to over a week under the influence of a high pressure system. Warm moist air from the southern latitudes moves northward on the west side of the high pressure center. (Remember, winds are southerly in the western portion of a high.) As the air travels northward, it encounters colder water and cools from below. If the sky is cloudless or only has high thin cirrus type cloud, a lot of heat is radiated away from this layer of moist air, and it may cool to the point where the water vapor condenses into droplets. Fog has formed.

Once fog forms, it continues its journey around the high, swinging down along the British Columbia coast southward to Washington on the northerly winds found in the eastern portion of the high. Additionally, especially in summer, the colder water lying along the coast may help form local sea fog. This, in turn, may join with the fog coming around the north end of the high. A vast area of ocean can be covered in a short period of time by this process as can be seen in the sequence of pictures described above. Sea fog will dissipate when the pressure pattern changes. If winds start blowing offshore, the dry continental air will push the fog farther to sea. If the high moves on or dissipates, more cloudiness is likely, which helps retard the loss of heat from the moist air layer next to the sea. Also, the temperature inversion aloft disappears, allowing the moisture in the air to mix to a higher level and reduce the humidity. Winds may increase with an approaching low pressure system. All of these processes help dissipate the fog or prevent its formation

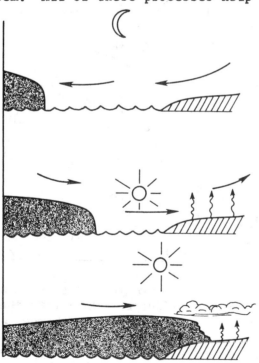

Nighttime:
Fog bank pushed offshore by land breezes.

After sunrise:
Land heats up and onshore breeze starts. The rising air from the land is replaced by air blown in from the ocean. The fog bank moves toward the shore.

Later...
Fog bank reaches shore but may break up into a layer of stratus farther inland.

Figure 5-14   Sea fog offshore may come ashore during the day due to the sea breeze caused by heating of the land.

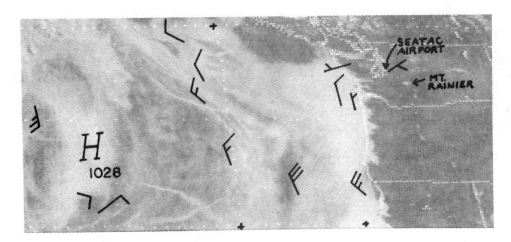

8 AM June 17

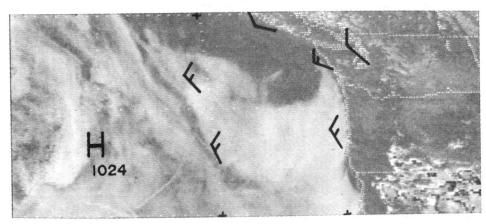

4 PM June 17

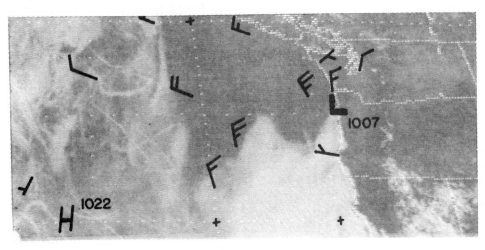

8 AM June 18

Figure 5-15 Weather satellite pictures of clouds over the ocean off the Pacific Northwest coast. The sequence shows how variable the area covered by fog can be. TOP PICTURE (8:00 a.m. PST 17 June). Fog (white area) lies along the coast. Winds offshore are northerly around the 1028 mb high. MIDDLE PICTURE (4:00 p.m. PST 17 June). Northerly winds offshore push the sea fog southward, expanding the clear area (dark part of picture) offshore. The streaks in the lower left of this picture are contrails from ship exhausts. BOTTOM PICTURE (8:00 a.m. PST 18 June). The entire area off Washington is now clear of fog. A weak low (1007 mb) has moved northward due to very strong heating originating in southern Oregon and California. The high has moved southward, but winds remain northerly off the coast.

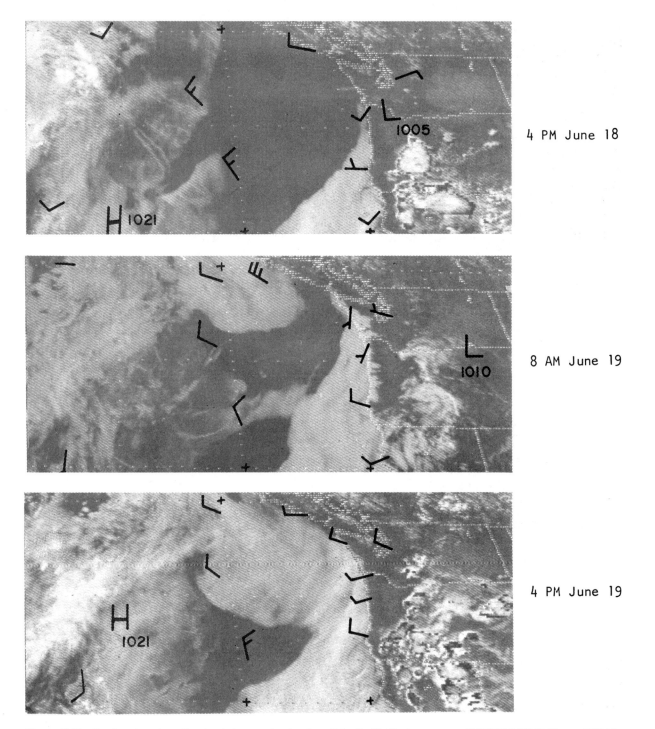

Figure 5-16 Continuation of satellite cloud pictures showing fog off the Pacific Northwest coast. TOP PICTURE (4:00 p.m. PST 18 June). The sea fog makes its greatest retreat southward. The "heat" low has started to move inland near Portland, and winds along the coast become more westerly or southwesterly. The fog starts to move northward again. Note that the high has weakened to 1021 mb and has moved farther south, but winds around the high continue northerly off the coast. A new patch of fog (fuzzy white area) has started to form southwest of Vancouver Island, MIDDLE PICTURE (8a.m. PST 19 June). The low has moved east of the Cascades into northeastern Oregon, and the high has moved even farther south out of the picture. Winds along the Washington coast are southerly, causing the fog to move up the coast even farther. Complicating the matter, the new fog patch forming off Vancouver Island moves southeastward due to northerly winds offshore. BOTTOM PICTURE (3:00 p.m. PST 19 June). The high pressure center has again moved northward. The sea fog southwest of Vancouver Island has joined with that along the Washington coast.

in the first place. A falling barometer, therefore, is a good indicator that the pressure pattern is changing and that the fog may be on its way out.

Fog that results from precipitation normally will last less than a day unless high pressure establishes itself after the rain or drizzle has fallen. If the winds die down after precipitation has occurred, and the skies clear off at night, sea fog may result. Extensive areas of fog are often found in the warm air sector between warm and cold fronts, but local fog can also occur with occluded fronts or cold fronts. These fogs will be short-lived, however.

Another type of fog -- steam fog or "sea smoke" -- can be found over the water when the air is very much cooler than the water, say by 15 to 20°F or so. The fog rises in whispy streamers off the water much like steam rising off boiling soup and may extend several hundred feet into the air. This fog seldom becomes as dense as the sea fog we have been discussing. It is not common off the Washington coast.

Chapter 6
SAILBOAT TACTICS AND WEATHER

The weather phenomena discussed earlier are uncomfortable or even dangerous situations for all small boat operators. In this chapter we look at some of the principles of small scale weather that are not necessarily bad weather situations, but which may determine whether or not the best possible progress can be made under sail. It must be pointed out that nobody knows what the winds and currents are like in every location under all kinds of weather patterns. What we will attempt here is to give some general guidelines that you might find useful and list some local rules formulated by longtime sailors. Finally, some thoughts on open-ocean sailing will be discussed, but by no means should these be considered to be the total sum of knowledge needed for an ocean passage.

Small Scale Windflow Principles

Windflow patterns range in scale from the molecular to global. Molecular motion is not of importance to sailboaters since this is simply the random motion of air molecules on a microscopic scale. On the other hand, very large windflow patterns are significant, such as the prevailing wind pattern around a high pressure system covering the eastern Pacific Ocean. The large scale weather pattern affects the local weather patterns imbedded within it, and it is these small local effects -- often less than a couple of miles wide -- that combine with the large scale effects to determine the actual conditions we observe.

Only by actual experience under various weather conditions will it be possible to correlate what is observed with information from another site. For example, knowing that the winds are northerly along Puget Sound outside Elliott Bay (from Weather Radio KHB-60 observations at West Point and Alki Point, discussed in Chapter 7), what are the expected winds in the unreported area along the north shore approaching Elliott Bay, just east of West Point, along Magnolia Bluff? There are no general rules that will answer questions like this 100% of the time. Far too many factors contribute to the winds on such a local scale. But the task of answering questions like this is not at all hopeless. Although there is no substitute for local knowledge when it comes to specific questions, we can go far with an educated guess, based on the shape of the shoreline and a few general principles.

Air Stability and Wind Gusts

In Chapter 4, air stability was shown to be very important in determining whether or not winds will be gusty. Winds will be gusty when the air can mix vertically as well as horizontally, which has the effect of transferring stronger, often veered, winds aloft momentarily to the surface. Winds are not very gusty when the air is stable and tends to move along in layers like slippery plastic playing card stacked on each other.

SIGNS OF UNSTABLE AIR:

1. Cumulus clouds.
2. Good visibility except in showers (if any).
3. Smoke plumes waft up and down and disperse; smoke does not drift in smooth layers.

SIGNS OF STABLE AIR:

1. Stratus or fog.
2. Poor visibility caused by pollutants and haze trapped in the lower levels of the atmosphere.
3. Smoke in layers. Smoke from chimneys on shore will drift in layers and be slow to disperse. In the Strait of Juan de Fuca, exhaust smoke from ships may be seen in layers the length of the Strait.

Fronts and lows moving into western Washington bring unstable conditions. Highs usually bring more stable air, but gusty winds can still occur during daytime due to thermal currents in the air created by solar heating. Also, if the air is colder than the water, heating from below will create thermal currents. As mentioned in Chapter 4, the sea breeze often reaches near gale to gale force in the eastern end of the Strait of Juan de Fuca, particularly in the evening and nighttime hours. These winds are very gusty and associated with intense heating in eastern Washington. It takes a while for the mass of air in the Strait to get moving, but once it does, the air continues to move through the Strait toward the Cascades. The northerly winds in Puget Sound are also affected by sea breezes. The area south of Olympia, having much less water coverage than areas to the north, heats up much more by late afternoon, which in turn sets up a sea breeze in Puget Sound. Fifteen to 20-knot northerly winds are typical, but they quickly die down after sunset.

Obstructions to the flow of air, such as islands and points of land, disrupt its movement. Expect to find gustier winds on the downwind side of such barriers as illustrated in Fig. 6-1. The stronger the wind and the more unstable the air, the gustier the winds will be.

Gale and storm force winds accompanying strong fronts and deep lows may be particularly hazardous because of wind gusts. As a very general guideline, if the average wind is less than about 30 knots, wind gusts may occasionally be 100% higher than this (a factor of 2 stronger), depending on stability of the air and location of obstructions. For average winds above 30 knots, wind gusts are more likely to be around 25% to 50% higher than the average. Table 6-1 shows actual examples of wind gusts measured at an instrumented site along the Oregon coast that were associated with the passage of a strong occluded front.

The gusty winds prior to the frontal arrival varied from 5 to 20 kn over the 3-min period and veered with each peak gust. Once the front passed by, winds became gustier, varying in speed from 31 to 68 knots. The direction from which the gusts blew also veered with each peak speed. In general when the the air is unstable, such as behind a cold front, wind gusts tend to veer from the average wind direction. The amount of veering may be more than 45° to the right of the average wind direction (when you face the wind). Winds in more stable air, such as behind a warm front or when the air is cooled from below by the earth or water, will not be nearly as gusty, and the change in wind direction will be small. Nearby terrain will also affect the gustiness due to "mechanical" mixing of the air as it blows over a rough surface.

Needless to say, conditions would be very unpleasant under sail in the example from Table 6-1 after the front passed by. The force of the wind is proportional to the square of its speed. In the example, momentary forces some 4 to 5 times greater are exerted during the 68-kn gusts compared to the 30-kn lulls. It is no mystery that sailboats can be knocked down in conditions like these.

On a long beat to windward in gusty winds there will often be a favored tack that optimizes your windward progress. The effect occurs because, as seen in Table 6-1, gust winds often shift in a consistent direction. Away from any strong influences of the local terrain, one might expect the gust winds to veer, in which case the starboard tack is favored, as illustrated in Fig. 6-2.

In some circumstances, gusts can occur consistently at the successive passage of each of a series of cumulus clouds. In this special case, a downburst gust could be anticipated at the approach of the next cloud. The key point here is not so much to predict theoretically the direction of the gust shift, but rather to note the direction of the gust, whatever it is, see if it repeats, and sail accordingly.

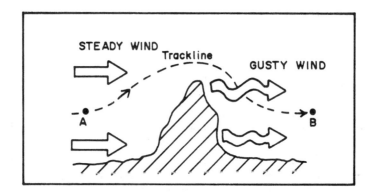

Figure 6-1 Gusty winds occur downwind from a point of land or an island. Wind at A is steadier than at B. This is a top view of the headland.

Table 6-1   WIND GUSTS WITH THE PASSAGE OF A STRONG OCCLUDED FRONT

Elapsed Time	One Hour Before Passage		Just After Passage	
	Instantaneous Wind	Shift Relative to Average Wind	Instantaneous Wind	Shift Relative to Average Wind
0.0 min.	E at 8 kn.	back	SSW at 64 kn.	veer
0.5 min.	SE at 20 kn.	veer	SSE at 31 kn.	back
1.0 min.	E at 7 kn.	back	SSW at 68 kn.	veer
1.5 min.	SE at 17 kn.	veer	SSE at 34 kn.	back
2.0 min.	ENE at 5 kn.	back	SW at 61 kn.	veer
2.5 min.	ESE at 12 kn.	no shift	S at 35 kn.	no shift
3.0 min.	ENE at 5 kn.	back	SSW at 57 kn.	veer
Average wind over 3 min. period was:	ESE at 10 kn.		S at 44 kn.	

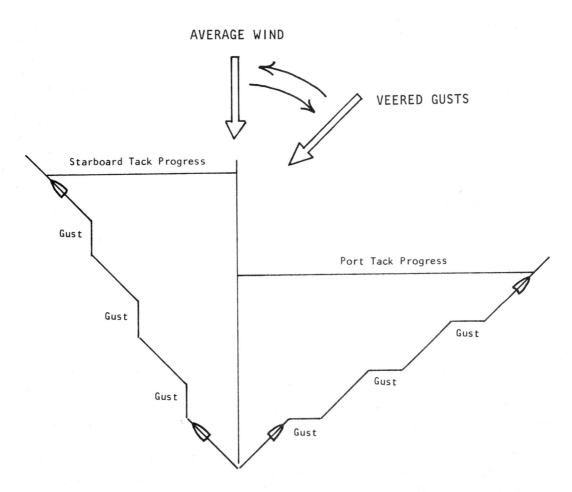

Figure 6-2 The advantage of sailing to windward on the "lifted" tack. Generally, a sequence of gusts will shift in the same direction, which means there is a favored tack for making maximum progress to windward in gusty winds.

Wind Around Islands

The windflow on the downwind side of an island is not likely to be the same as on the upwind side. The islands in western Washington are not conical, so the simplified illustration of this in Fig. 6-3 (based on a number of measurements around an isolated hill) can only be used as a rough guide. Generally, the stronger the prevailing wind, the greater the changes in airflow around the island and the further these changes will be felt downwind. Also, near the tips of the island, wind speeds may be greater than they are further offshore. Be prepared for unusual windflow patterns around islands in strong winds.

The type of flow shown in Fig. 6-3, by the way, applies very much to the Hawaiian Islands where the prevailing NE trade winds curve around each island. This effect in the Hawaiian Islands is illustrated nicely in a diagram on the back of NOAA's "Marine Weather Services Chart" for the Hawaiian Islands, available from the National Weather Service and some marine supply stores.

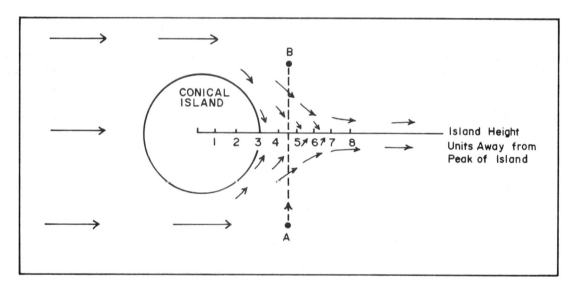

Figure 6-3 Windflow around an island having a cone shape. The wind on the downwind side curves around the island. Wind speed rapidly drops off at the center line in traveling from A to B, and the direction of the wind reverses also. Wind speeds pick up as the edge of the island is approached. At a distance of about 7 or 8 times the height of the island, winds again are similar to what they are on the windward side. Winds may be calm on the lee side if close passage is made.

Channeling and Wind in Passes

Airflow that encounters a constriction between land masses will increase its speed through the constricted area. Wind speeds in this area can be double what they are a few miles either side of the constriction. Figure 6-4 illustrates some examples where you can expect increased wind speeds.

Wind in passes is an amplification of the wind by channeling, discussed in Chapter 4. In constricted areas, the wind follows the shape of its boundaries. A steady north wind in central Puget Sound will generally back to the northwest at the south tip of Whidbey Island as the shape of the Sound curves off to the northwest. Sailing farther north, the northwest wind will back on around to the west as you approach Point Wilson and enter influences of the westerlies from the Strait that are channeled around into the Sound.

On a smaller scale, this type of channeling can provide a dramatic tactical advantage when sailing through a long curved passage, such as Colvos Passage, west of Vashon Island. The trick is to sail point to point, tacking just off the point, where the wind curves back toward the shore. This provides a "lifted" tack both on the approach and departure from the point, as shown in Fig. 6-5. The trick might be used to advantage on any point of land where you expect the wind to follow the curve out and around the point.

Slope Winds

When the general windflow over western Washington is "light and variable," extremely local and small scale windflow patterns will be observed. The most notable of these is the slope wind caused by cooling of the ground at night. As the hillsides cool, air next to the ground on the slopes cools faster than the air over the water. The result is that the denser air on the slopes slides down the hills as a gentle wind, usually less than 10 knots. If fog has formed

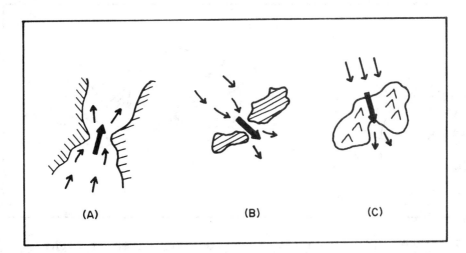

Figure 6-4 Constrictions in flow of air increases wind speed.
    (A) Channel narrowed by land masses.
    (B) Channel between two islands.
    (C) Bay downwind from a pass between hills or mountains.

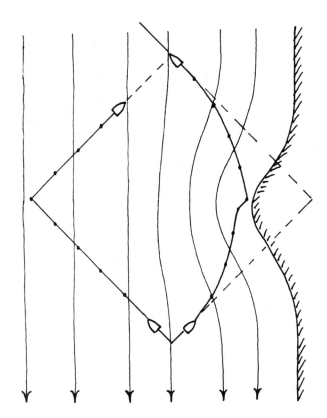

Figure 6-5 The advantage of sailing point to point. The boat that tacked at the point was "lifted" on its approach and departure. The wind remains at about 45° to the boat's heading at all times, on each tack.

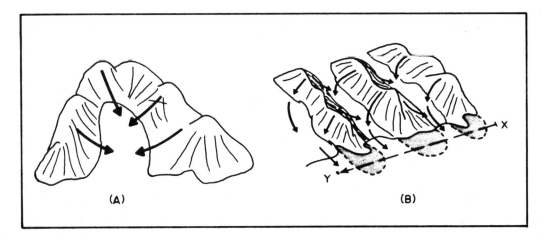

Figure 6-6 Slope winds bring light breezes over water at night.
(A) The direction the wind blows is determined by the orientation of the terrain. (B) Slope winds flow down into valleys and out over the water between the hills. In a transit from X to Y, areas of very little or no wind would be found in the shaded areas. Winds would be strongest between the hills.

over land, particularly in the valleys, it will be carried out over the water by these breezes. Figure 6-6 shows the principle behind slope winds.

Slope winds seldom extend more than a mile offshore, and often much less than that. The flow of air behaves much like the cold air that runs out of an opened refrigerator down onto the floor. This process may be partially responsible for the light evening winds off the east shore of northern Puget Sound that are reported by longtime sailors in this area.

Winds Near Thunderstorms

Lightning and gusty winds are the main threats from thunderstorms in western Washington, but in other parts of the U.S., heavy showers of damaging hail also have to be considered. To be safe from lightning, the mast of a boat must be properly grounded so that a strike can be conducted down the mast and into the water without passing through the people on board or through the hull of the boat itself. Lightning can literally blow a large hole in the bottom of a vessel.

Winds associated with thunderstorms are often gusty and strong because of tremendous downdrafts in parts of the storm that continue all the way to the ground (or water) where they spread out in all directions from the thunderhead.

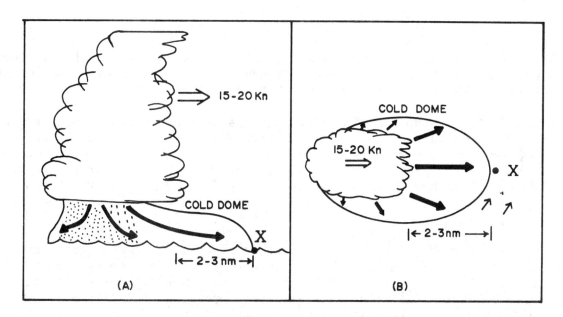

Figure 6-7 Windflow beneath a thunderstorm. (A) Side view: cold air sinks down through the thunderhead and flows out over the earth's surface. Movement of the storm, in this case 15-20 kn from left to right, causes the dome of cold air to assume an oblong shape. Very choppy waves may be created.
(B) Top view: storm is moving left to right at 15-20 kn, and strongest winds are ahead of the storm toward its direction of movement. Winds outside the cold dome are not the same as inside the dome. At point X, winds will veer (change clockwise in direction) from SW to W once the edge of the dome passes, which will only take a few minutes at most. Air temperatures may drop a few °F to over 20°F.

Wind gusts of over 50 knots are possible. The higher the thunderhead extends into the sky, the stronger this wind will be. Figure 6-7 shows how the downburst fans out from the storm. A common sense rule when it comes to thunderstorms is that it if looks unusually threatening, it very likely is.

Once a thunderhead gets its "anvil top" it has reached maximum growth and will start weakening. It is possible to have clusters of thunderstorms together, which complicates the windflow pattern shown in the diagram. The cold dome may extend ahead of violent thunderstorms by some 15-20 nm, but this would indeed be a rare event for western Washington.

Local Wind Effects in Puget Sound of Interest to Sailors

This section summarizes a few local wind features that are specific to Puget Sound. Several of these are discussed in earlier parts of this book -- the convergence zone and the thermal (sea breeze) northerly, for example, are well documented effects based on official weather observations. Other features listed here, however, are not based on official records, but rather on the extensive observations of sailors in the area. They represent the best guesses that experienced sailors might make about the behavior of the wind in certain areas. Nothing is guaranteed. But these things have happened often enough that they are remembered and anticipated.

(1) WIND FOLLOWS THE SHAPE OF THE SOUND. Look at a chart and the shape of the Sound to guess if, for example, a "southerly" is likely to be more southeasterly or southwesterly in a given region. Or, if the forecast specifically calls for a northeasterly, as another example, look for the winds to be steadier or stronger in an area that opens up to the northeast, and so forth.

(2) LOOK OUT FOR WIND SHADOWS. Steady winds in the center of the Sound may be shadowed inside bays, close to the leeward side of hills along the shore, or just around the leeward corner of an island. For example, winds can be quite weak, even in the presence of strong mid-sound northerlies, at the south tips of Whidbey or Vashon Islands. Sailors often duck into bays or coves to avoid adverse mid sound currents. In doing this, one must balance out the chance of losing wind.

(3) THERMAL NORTHERLY. The wind in the Sound builds from the north as the land to the south heats. The time it starts depends on the temperature and sky cover. In the summer it can start in mid morning, but generally it is an afternoon effect, peaking in late afternoon and going out like a light with the sunset. This wind can reverse a morning southerly or dramatically enhance a northerly.

(4) NORTHERLY FILLS FROM THE NORTH AND THE WEST. The north wind will usually fill from the north, meaning the line between a south wind (or no wind) and the north wind generally moves down the Sound from the north. This is generally true with the thermal northerly -- or any northerly that is filling in the Sound with an associated westerly filling in the Strait of Juan de Fuca. If you must guess which side of the Sound will get the north wind first, it would be the west side. Anticipating a northerly when in mid Sound, head west.

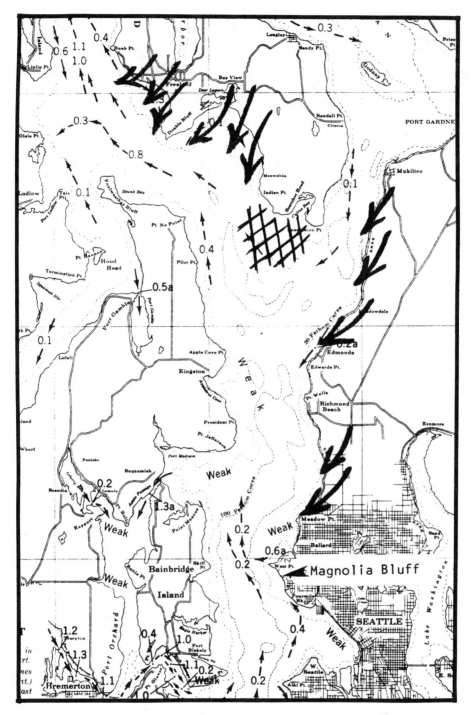

Figure 6-8 Areas in Puget Sound with consistent slope or drainage winds. Look for light NE winds in these areas in the evening after the thermal northerly dies off. Although light winds may persist here throughout the night, just after sunset and just before sunrise would be the optimum time to find them. These weak winds will not be noticed when there are stronger winds in the Sound. The cross-hatched area off Cultus Bay is a wind shadow for these winds. The chart used in this figure is reduced from the "Tidal Current Charts for the Northern Part of Puget Sound," referenced in Chapter 2.

(5) WHEN THE WIND DIES AT SUNSET, HEAD EAST. If the wind in mid Sound dies at sunset, in the northern part of the Sound you can often find light northeasterly wind along the eastern shore of the Sound. These areas are shown in Fig. 6-8. Summer dawns will often reveal an entire racing fleet lined up along these shores ghosting along in the only wind around.

(6) CONVERGENCE ZONE. If the wind is from the north in the north Sound and from the south in the south Sound, then it has to be weak and variable where the two winds meet. This often occurs near the southern end of Whidbey Island. There are no tricks for getting through this calm area, but keep in mind the currents if you expect a convergence zone. Make way toward the area where the currents will do the least damage. The convergence zone is discussed in detail in Chapter 4.

(7) MAGNOLIA BLUFF. To complete the example mentioned at the beginning of this chapter, we consider a very local effect just east of West Point, under Magnolia Bluff. Here one might guess from the shape of the shoreline that a mid Sound northerly might wrap around West Point in toward Elliott Bay to form a northwesterly in this area. But on the contrary, the wind in this area, in these conditions, is typically from the northeast. Apparently the overall shape of the hills in this area is more important than the shoreline profile. The north wind must ride up over the more gentle slope of the hill from the north and be bent to the west as it goes, so it ends up spilling down off the Bluff from the northeast when it reaches the water.

Knowledge of wind effects like those mentioned above can make your cruising sailing more efficient and your racing sailing more competitive.

Ocean Sailing

Unless you have a radiofacsimile recorder aboard to copy weather charts, ocean sailing involves a high degree of uncertainty and luck when it comes to the weather. Weather information you get at sea is rather limited without radiofacsimile, and because of the slow speed of most sailboats, it is very difficult to maneuver out of the way of a storm, even with 24 hours' warning. A boat venturing out to sea must truly be a seaworthy craft, and personnel should be capable of handling her under adverse conditions should they arise.

The positions of troughs and ridges are of utmost importance (see Chapter 8). An elongation in a particular direction of the isobars around a low pressure center is called a trough. Almost all lows have one or more troughs, most of which will point toward the SE to SW from the low center. East of the troughline (line connecting the sharpest bend in the isobars) winds are southwesterly to southerly, depending on the orientation of the trough. West of the troughline, winds will be southwesterly to northwesterly, again depending on the orientation of the trough. The air mass will also be cooler and more unstable on this side of the trough, and so showery weather and gusty winds are more likely. The more pointed the isobars are that make up the trough, the stronger the winds and the worse the weather. Wind directions change rapidly from one side of the trough to the other in this case. Weather fronts are found in troughs, but not all troughs have weather fronts. Words, such as "deep trough," "sharp trough," "strong trough," in the weather bulletins should alert you to the fact that the trough may pack quite a punch.

An elongation of the isobars in a certain direction around a high pressure center is called a ridge. The ridgeline, like the troughline, extends through the greatest bend in the isobars. Ridges usually point in a NE to NW direction, and it is often observed that another ridge south of the high center points in a southerly direction. The change in wind direction in crossing the ridgeline is usually not as marked as across a troughline. Wind directions east of the ridgeline vary from northwesterly to northeasterly, depending on orientation of the ridge. On the west side of the ridgeline, winds are southeasterly to southwesterly. Weather conditions are likely to be worse on the west side of the ridgeline since weather fronts move in a general west to east direction. East of the ridgeline, the weather may vary from clear skies to heavy fog and low clouds. If the ridge is stationary, it will usually prevent weather fronts from passing through it, or it will weaken those that do make it through. Figure 6-9 shows the winds around troughs and ridges.

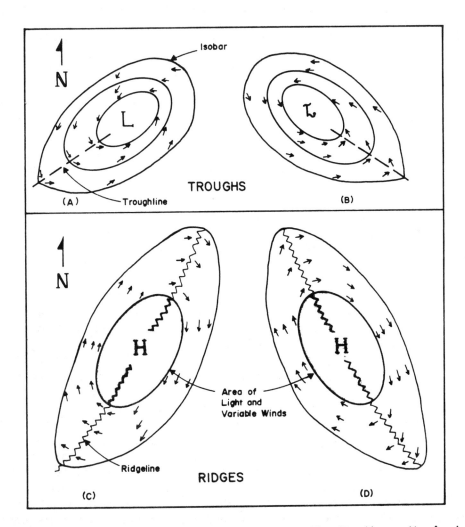

Figure 6-9 Winds around troughs and ridges in the Northern Hemisphere.
(A) Trough oriented SW.   (B) Trough oriented SE.
(C) Ridges oriented NE and SW.   (D) Ridges oriented NW and SE.

The shape of the isobars determines not only the direction in which the trough or ridge is oriented but also the intensity of wind and changes in wind direction. Without weather maps, you will not likely know too much about the shape of the trough or ridge. In general, when southbound off the coast, best progress is likely on the west side of a trough or the east side of a ridge. The converse is true when sailing northward. The problems come when the troughs or ridges are not stationary, which may soon put you on the wrong side.

Winds are light and variable or even calm within about 200 nm of most high pressure centers over the ocean. If the high is stationary or nearly so, it is best not to approach the high center any closer than 2 to 4 mb unless you want to do some motoring. For example, if the high has a central pressure of 1029 mb, sailing into pressures higher than about 1025-1027 mb may very well put you in light to calm winds. Remember that positions of high pressure centers as given in weather bulletins may be in error by 100-200 nm (or more) because of the lack of weather observations available to locate the exact center. Unlike low pressure centers, it is often not possible to identify high pressure centers using satellite cloud pictures.

Winds are also light in the region between two low centers because there is not much pressure difference between the centers, which makes the pressure gradients weak. Caution must be used, however, when considering light winds between two low pressure centers. Sometimes winds will be very strong within several hundred miles of each center, and it is not until you are outside of these circulation areas that the winds are weak. This is not a problem with high pressure centers as long as a trough or low center is not located between

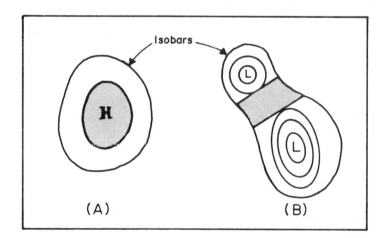

Figure 6-10 Locations of light and variable winds (shaded areas). (A) "Inside the last isobar" around a high center. Isobars on weather maps are drawn at 4-mb intervals: for example, 1016, 1020, 1024, etc. As a general rule, you will enter the region of light and variable winds when your barometer reads a pressure within 4 mb of the reported pressure for the center of the high. (B) Between two low centers as long as you remain out of the areas of strong circulation around each low, which depends on the depth of the lows involved. The deeper the lows, the larger the circulation areas usually are, so by passing halfway between the two centers you will miss the stormy area. High swells are a possibility in this "null" area, but sometimes the ocean is nearly calm.

the two, which is often the case. The area halfway between two high centers is a prime location for the formation of weather fronts. Figure 6-10 shows the areas where winds are likely to be weak (less than about 10 kn and possibly calm).

Buys Ballot's Law

The direction in which a low pressure center and a high pressure center lie can be determined on the open sea by simply using the direction of the wind. In the Northern Hemisphere, if you stand with the true wind blowing on your back, lower barometric pressure lies to your left and higher pressure toward your right as shown in Figure 6-11.

Remember that the apparent wind direction from a moving vessel -- made up of the true wind and the wind created by the motion of the boat -- can be significantly different from the true wind direction of interest here, especially when reaching across the wind. If in doubt about this correction, you can always stop the boat to check the true wind direction.

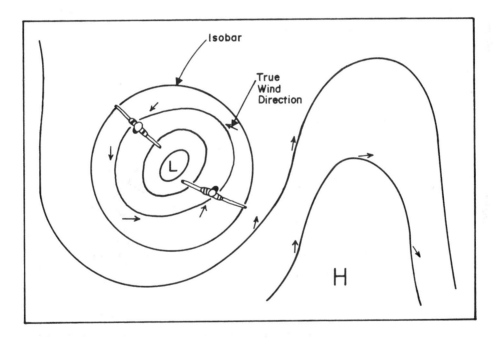

Figure 6-11 Buys Ballot's law. With your back to the true wind in the Northern Hemisphere, lower pressure is to your left and higher pressure is to the right. Winds around a low center typically blow at a 10-20° angle to the isobars toward lower pressure, so the location of the low center would be forward of your left arm by 10-20°. The more circular the isobars are around the low, the better this approximation works. Nonetheless, regardless of the shape of the isobars, lower pressure will be to your left. Similarly, winds around a high blow at a 10-20° angle to the isobars away from the high center. The high center would be located 10-20° behind your right arm. Buys Ballot's law does not work in protected waters, only on the open sea. This relationship between wind and pressure was formulated by the Dutch meteorologist, Buys Ballot, in 1857.

High Seas Weather Information

Weather information for the high seas (more than 250 nm offshore) is available from Coast Guard Communications Station NMC and commercial radio stations KMI and KFS, all of which are in or near San Francisco. The weather bulletins are prepared by the National Weather Service Forecast Office at Redwood City, California. Radio stations in Hawaii and Alaska also broadcast high seas information, but we will limit our discussion here to those on the U.S. West Coast. KFS only transmits via radiotelegraph (Morse code); KMI only on radiotelephone (voice); NMC on both radiotelegraph and radiotelephone. Complete radio schedules and frequencies are published in <u>Worldwide Marine Weather Broadcasts</u> compiled by NOAA and the Navy. This publication, which is updated annually, is available from the Superintendent of Documents, U.S. Government Printing Office, Washington, D.C. 20402 or at local U.S. Government Printing Office bookstores.

In addition to the three stations just listed, storm information and locations of major highs and lows are available from WMV and WWVH, the two Bureau of Standards' time stations used for navigation. The schedules are in the publication just mentioned. High Seas weather services are discussed further in Chapter 7.

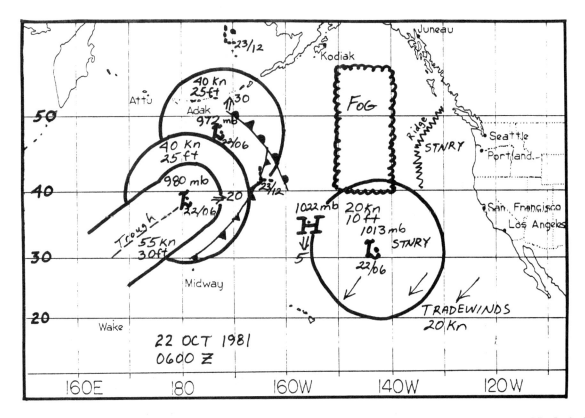

Figure 6-12 Plot of weather information in high seas bulletin of 22 October 1981 at 0600 GMT. Data below 20N are not plotted. Present positions of highs and lows are shown at 22/06 (22 Oct./0600 GMT). Predicted locations are shown with dashed letters at 23/12 (23 Oct./1200 GMT). STNRY means stationary.

Many people find tape recorders very handy for copying the broadcasts from any of these stations since there is a lot of information given out and reception is sometimes poor. One way to make the best use of the information is to plot it on a small chartlet, making a new plot for every new broadcast. Thus, a series of weather maps will be complied, and trends in the pattern can be picked out. It is best to copy at least two broadcasts a day as the minimum needed for tracking the weather. An actual high seas bulletin is shown below, and the information is plotted in Figure 6-12.

A shortwave receiver capable of receiving these broadcasts should be considered the minimum radio equipment for any ocean or coastal cruise.

NATIONAL WEATHER SERVICE SAN FRANCISCO. HIGH SEAS WEATHER 220600 GMT OCTOBER 1981. FORECAST POSITIONS 231200 GMT..... WARNINGS AND FORECASTS: STORM 980 MB 39N 178W MOVING EAST 20 KT. FORECAST POSITION 43N 163W. A TROUGH LINE EXTENDS TO 37N 178E 35N 175E 31N 167E. WIND 55 KT SEAS 30 FT WITHIN 250 MI OF CENTER AND TROUGH. GALE WINDS TO 40 KT SEAS 25 FT ELSEWHERE WITHIN 500 MI OF THE CENTER..... GALE 972 MB 47N 172W MOVING NORTH NORTHEAST 30 KT. FORECAST POSITION 59N 170W. OCCLUDED FRONT FROM 51N 172W TO 45N 164W THEN WARM FRONT TO 40N 160W AND COLD FRONT TO 40N 165W 30N 178W. WIND 40 KT SEAS 25 FT WITHIN 500 MI OF THE CENTER..... AREAS OF DENSE FOG BETWEEN 40N AND 55N AND BETWEEN 140W AND 150W..... GALE WINDS 35 KT SEAS 18 FT ARE EXPECTED TO DEVELOP IN THE GULF OF THEUANTEPEC ABOUT 230600Z..... SYNOPSIS: LOW 1013 MB 32N 143W NEARLY STATIONARY. WIND 20 KT SEAS 10 FT WITHIN 500 MI OF CENTER..... STATIONARY RIDGE THROUGH 52N 130W, 45N 135W, 40N 134W..... HIGH 1022 MB 35N 155W MOVING SOUTH 5 KT..... TRADEWINDS 20 KT, SEAS 10 FT, SOUTH OF 38N AND WEST OF 120 W..... SCATTERED THUNDERSTORMS 100 MI OF A LINE 5N 82W, 10N 90W, 10N 95W, 8N 105W, 8N 121W..... SCATTERED THUNDERSTORMS WITHIN A 60 MILE RADIUS OF 22N 116W, ALSO WITHIN A 60 MILE RADIUS OF 15N 124W..... SCATTERED THUNDERSTORMS BETWEEN 6N AND 13N AND BETWEEN 125W AND 131W..... SCATTERED THUNDERSTORMS WITHIN 60 MI OF A LINE FROM 9N 131W TO 10N 135W.

Chapter 7
FORECASTS, WARNINGS, AND SOURCES OF WEATHER INFORMATION

Purposes of Forecasts

What is the purpose of a weather forecast? The answer seems immediately obvious for we say right away that it is supposed to tell us what the weather will be in the future at some specific location. However, of necessity, the forecast must cover a large geographical area, such as the Strait of Juan de Fuca, because it would otherwise be too lengthy.

If the wind is forecasted to be easterly at 20 to 30 knots and we are just off Ediz Hook near Port Angeles in light winds less than 5 knots during the time the forecast is supposed to verify, the forecast is not accurate to us. Our friend who is in the middle of the Strait radios us and says that it is blowing 25 knots from the southeast out where he is. The forecast is an accurate one to him. Wind speed forecasts are always for mid-channel areas least influenced by land. Forecasts seldom will give you the wind you observe in areas highly affected by terrain. This will be a problem for some time to come because the state of the forecasting art and the means of disseminating more detailed forecasts are not yet capable of giving more detail. When you consider the complexity of the terrain in northwestern Washington, you can appreciate the monumental difficulties there would be in providing a forecast for every channel, bay, and inlet. You can only gain knowledge about the local effects from actual experience or from those who have been there. It also helps to know weather basics so you can relate what you experience to the principles behind the cause of the weather.

Using Forecasts

One of the most important things you can do when using forecasts is to determine the trend; that is, are conditions expected to improve, stay about the same, or worsen? A simple example would be if you were on the leeward side of an island protected from strong southerly winds, but not from northerly ones. The forecast predicts northerly winds to 30 knots by the afternoon, which means a worsening trend in your situation. You should now be on the alert to the possibility of having to move, and should fix in your mind what conditions pose unacceptable risks to you. If you are in a rowboat, you will want to move before the winds come up because it is very hard to row against a 30 knot wind. Yet, if you are in a 36-foot power boat, you can easily make headway and may want to wait a while longer to see if the wind does in fact materialize. It is important to keep in mind what wind directions and speeds are important for your activities at a given location so that you can plan accordingly.

Rain or snow are usually not too hazardous, but they certainly cause much discomfort if you are not properly prepared. The forecast contains a percentage chance of precipitation based on the forecaster's estimate of the odds of it occurring with a particular weather pattern. For the Seattle area, a 90% chance means that in nine cases out of ten that .01 inch or more of precipitation will fall in the rain gauge at SEATAC Airport (the official verification point for Seattle) during the period for which the forecast is valid. Since the probability is high for SEATAC, we can also assume it will be high for Edmonds because this is still close to SEATAC as far as large scale weather patterns are concerned. If an afternoon shower comes up and we get soaked, but not one drop falls in the bucket at SEATAC, the forecast was good for us, but not good for people near the airport. Officially, in fact, the forecast did not verify. Precipitation forecasts are right about 85% of the time (up to about 24 hours) and in error about 15% of the time. The old Boy Scout maxim, "Always be prepared," very much applies to boating and your use of forecasts.

Since the forecast is nothing more than an estimate of future conditions, you should look for weather signs either confirming or contradicting the forecast. As an example, if the forecast calls for increasing southerly winds of 20 to 25 knots in the Inland Waters by the afternoon as a front approaches, and you are presently experiencing light and variable winds (winds less than 5 knots and from various directions), the trend is for worsening conditions. You should observe, as time goes on, increasingly persistent and stronger winds from the south, increasing cloudiness as the front approaches, and falling pressure on your barometer if you have one.

What if the forecast is not correct? Suppose the forecasted trend is okay, but the timing is off. You will observe the above signs either sooner or later than what were indicated in the forecast. If the forecasted trend for worsening conditions is not correct, you are likely to continue experiencing weather very much like what it has been. If, on the other hand, the forecast indicates things are going to get better, and you observe that conditions are not improving, you should be prepared for the possibility of worsening conditions.

The marine forecasts for the Inland Waters and Strait of Juan de Fuca contain predictions about the wind, weather, and possibility of fog (when appropriate) plus any advisories or warnings for mariners. The coastal forecast (shore to 60 nm out), in addition to the items just mentioned, contains predictions about waves. Describing these might be likened to describing the height of trees in a large forest. There are a few that are very much taller than the rest, a large number more or less the same, and some that are very much shorter. Try to describe this in 10 to 12 words and you will see how difficult it is. The marine forecaster faces this problem by using significant wave heights (SWH), discussed in Chap. 2. From the value of the SWH you can make judgments about the range of wave heights you are likely to encounter. Table 2-1 in Chap. 2 should help with this.

When the forecaster is reasonably confident he can do so, he will separate the waves into two groups: wind waves (those caused by wind stress on the water) and swell. Predictions of significant wave height for each type will be given, for example: WIND WAVES 5 to 7 FT. WESTERLY SWELL 3 to 5 FT. Forecasts will usually have a range of significant wave heights, not just a single value. The values do not apply to heights of breakers on the beach or waves in shallow water, say less than about 150 ft. If the forecaster cannot separate the two types of waves, he will use the term "seas," which combines wind wave heights with swell heights into one value. (Mariners use this term synonymously for wind waves, so the meaning is different as used by the NWS.) Nonetheless, conditions under which the term is used for Washington coastal waters will usually be stormy, in which a 4 ft. swell, for example, would not be too important if 16 ft. wind waves are expected. As explained in Chapter 2, the "combined seas" (wind waves plus swell) would have a significant height of 16.5 ft., not 20 ft. The heights are not directly additive, that is, a 4 ft. swell and a 16 ft. wind wave do not combine to makes waves 20 ft. Rather, the square root of the sums of the squares of the individual heights is taken. In this case, the square root is taken of 4x4 plus 16x16, which is about 16.5.

As a mariner, we usually do not have to worry about this technical aspect. The important thing to remember is that when we hear a forecast for 10 ft. waves, we can reasonably expect (Table 2-1) an occasional one 90% higher or 19 ft. The most frequent height will be about 50% of the significant wave height or 5 ft. Quite a range, indeed!

As you can see by now, nothing is certain when it comes to using forecasts, but if you use them in terms of drawing conclusions about acceptable risks, you will be using them in a meaningful way.

Small Craft Advisory

This is used to alert boaters to weather or sea conditions, either presently occurring or expected to occur, that might be hazardous to small craft (vessel less than 65 feet in length). The threshold wind speed for which the National Weather Service will issue a Small Craft Advisory (SCA) is 21 knots with an upper limit of 33 knots. The Canadian Atmospheric Environment Service has 18 knots as its lower limit and 33 knots as the upper limit. Of course this is a wide range of wind speed and by looking at the wave tables in the Appendix, you will note that wave heights not only are going to be much higher at 33 knots than at 21 knots, but that the waves will get higher much quicker at the greater wind speed. SCAs are also issued if swells are expected to be 10 ft. or higher. An SCA for the Columbia River Bar or Grays Harbor Bar is issued if very steep or breaking waves are expected. Small Craft Advisories are sometimes ignored, which may not be prudent seamanship.

Gale Warning

This is issued to indicate that winds of 34 to 47 knots are expected. Some storms that affect our area produce gale force winds, but may, in a short time, produce winds much greater than 47 knots. You will note from the wave tables that the larger the body of water, the higher the waves will become for the same wind speed. You must evaluate very carefully the acceptable risks to you when you are faced with a gale warning. Remember, the forecaster is basing his judgement on a large input of data and is making his decision using the best of his skills.

Warning Dislpays and Explanations, reprinted from the U. S. Coast Pilot

NATIONAL WEATHER SERVICE COASTAL WARNING DISPLAYS

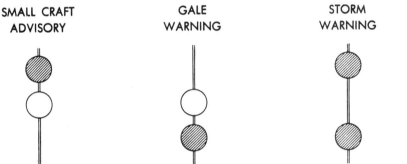

Note: Shaded area represents the color RED on flags and lights.

EXPLANATION OF DISPLAYS

Small Craft Advisory: One RED pennant displayed by day and a RED light ABOVE a WHITE light at night, to alert mariners to sustained (more than two hours) weather or sea conditions, either present or forecast, that might be hazardous to small boats. Mariners learning of a Small Craft Advisory are urged to determine immediately the reason by tuning their radios to the latest marine broadcasts. Decision as to the degree of hazard will be left up to the boatman, based on his experience and size and type of boat. The threshold conditions for the Small Craft Advisory are usually 18 knots of wind (less than 18 knots in some dangerous waters) or hazardous wave conditions.

Gale Warning: Two RED pennants displayed by day and a WHITE light ABOVE a RED light at night to indicate that winds within the range 34 to 47 knots are forecast for the area.

Storm Warning: A single square RED flag with a BLACK center displayed during daytime and two RED lights at night to indicate that winds 48 knots and above, no matter how high the speed, are forecast for the area. However, if the winds are associated with a tropical cyclone (hurricane) the STORM WARNING display indicates that winds within the range 48 to 63 knots are forecast.

Hurricane Warning: Displayed only in connection with a tropical cyclone (hurricane). Two square RED flags with BLACK centers displayed by day and a WHITE light between two RED lights at night to indicate that winds 64 knots and above are forecast for the area.

Note: A "HURRICANE WATCH" is an announcement issued by the National Weather Service via press and radio and television broadcasts whenever a tropical storm or hurricane becomes a threat to a coastal area. The "Hurricane Watch" announcement is not a warning, rather it indicates that the hurricane is near enough that everyone in the area covered by the "Watch" should listen to their radios for subsequent advisories and be ready to take precautionary action in case hurricane warnings are issued.

Note: A SPECIAL MARINE WARNING BULLETIN is issued whenever a severe local storm or strong wind of brief duration is imminent and is not covered by existing warnings or advisories. No visual displays will be used in connection with the Special Marine Warning Bulletin; boaters will be able to receive these special warnings by keeping tuned to a NOAA VHF-FM radio station or to Coast Guard and commercial radio stations that transmit marine weather information.

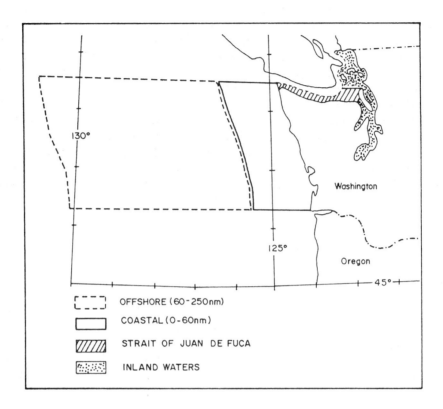

Figure 7-1 Marine forecast areas used by the National Weather Service for Washington. The NWS Forecast Office in Seattle makes forecasts for U.S. waters only, and the Canadian Atmospheric Environment Service in Vancouver, B.C., makes forecasts for adjoining Canadian waters only, which, because of local terrain effects on the weather, may result in different forecasts. Both U.S. and Canadian forecasts may be correct for their respective sides of the border.

Storm Warning

This is the highest level of marine warning (other than hurricane warnings issued in tropical climates) and is issued when the forecaster concludes that average winds of over 48 knots will occur. Unless you have a ship, acceptable risks in all open waters expect a pond do not exist under storm conditions, for to be caught in exposed waters miles from shore in a small boat in very high winds and seas will be a frightening experience -- perhaps a final one.

One of the worst storms of this century to hit the Pacific Northwest occurred on Columbus Day, October 12, 1962. Sustained winds over 50 knots with gusts to 87 knots (100 mph) roared over the Inland Waters and Strait of Juan de Fuca. Waves off Edmonds were averaging 5 to 6 feet with maximum heights around 10 ft. Another storm, the one of February 13, 1979, did not affect as large an area as the Columbus Day storm, but it did cause the Hood Canal Floating Bridge to break apart and sink due to 100 knot winds and high waves. Storm warnings had been issued well ahead of both storms.

Marine Forecast Areas

Forecasts and advisories or warnings are issued for the areas shown in Figure 7-1 by the National Weather Service for U.S. waters. The Atmospheric Environment Service of Canada issues similar warnings and forecasts for adjacent Canadian waters, and as much as possible, the forecasters involved coordinate their decisions so that the same category of warning is issued both by Canada and the U.S. There are cases, however, when winds in Canadian waters are different than those in U.S. waters, so the types of warnings issued can be different. This is especially true in the Strait of Juan de Fuca where it is not uncommon to actually have stronger winds on the Canadian side.

VHF-FM Continuous Weather Radio

VHF Weather Radio broadcasts are the most up to date and easiest to obtain sources of information. The letters VHF-FM are an abbreviation of the radio signal type: Very High Frequency - Frequency Modulated -- the frequencies used are above those you receive on a regular FM radio, and so a special radio is required. Most VHF radios for marine communications also include the proper weather frequencies. Four radio transmitters are operated by the National Weather Service (NWS) and one by the Atmospheric Environment Service (AES) of Canada for waters of western Washington and southern British Columbia. The NWS stations are KHB-60, KIH-36, WXM-62, and KEC-91. The AES station is CFA-240.

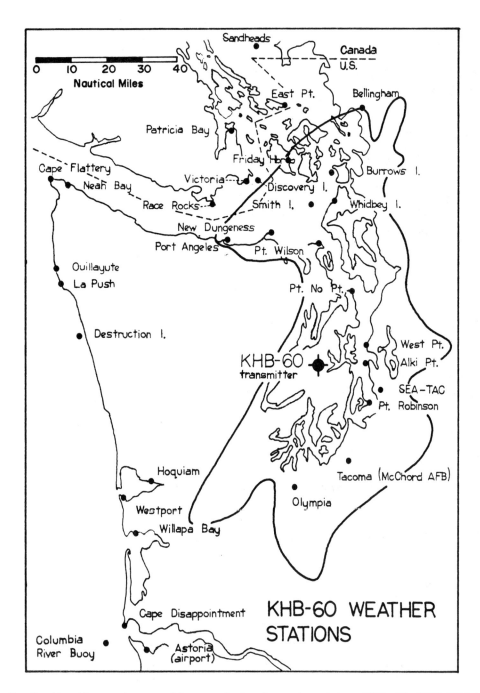

Figure 7-2 Weather observation points broadcast on KHB-60 VHF NOAA Weather Radio. Approximate area of reception shown by dark heavy line.

Weather Radio KHB-60

Frequency: 162.55 MHz ("Weather 1" or "WX1" on some VHF radios)
Transmitter location: Gold Mountain near Bremerton

Forecasts and weather observations are broadcast 24 hours a day. Contents of the transmission are:

1. Station Identification
2. Marine Forecasts
   a. Inland Waters
   b. Strait of Juan de Fuca
   c. Washington Coast (shore to out 60 nautical miles)
3. Weather Observations (See Figure 7-2)
4. Weather Summary, Seattle Forecast, Mountain Forecast
5. Western Washington Forecast
6. Eastern Washington Forecast
7. 6 - 10 Day Outlook
8. Seattle Forecast
9. Travelers Forecast, River Reports, Pass Reports, Avalanche Statements, other pertinent weather statements

Weather observation points for KHB-60 shown in Figure 7-2 are located at sites of varying distance from the water. Alki Point, for example, should give wind values more applicable to over-the-water than, say, SEATAC Airport. Not all stations take observations around the clock, so some reports are omitted from 5:00 PM to 6:00 AM PST, but since the NWS is in the process of automating most of the stations, this problem may eventually be eliminated. We will not get into the topic of hours of operation at each site since the situation is in a constant state of change. At times, due to communication problems or equipment failures, regularly scheduled observations are sometimes omitted on the VHF broadcast.

BROADCAST AND UPDATE TIMES:

The weather broadcast is continuous, 24 hours a day, but the information in the broadcasts does not change each time the broadcast is given. Although observations are taken hourly at many sites, the OBSERVATIONS are only updated every 3 hours on VHF radio (at 1:00, 4:00, 7:00, and 10:00 AM and PM, PST in the winter, and 2:00, 5:00, 8:00, and 11:00 AM and PM, PDT during the summer half of the year.) This is due to the fact that the weather reports have to be collected at the forecast office in Seattle and put on the recording that is transmitted from the VHF antenna site. The MARINE FORECASTS are updated at 2:30 and 8:30 AM and PM, PST in the winter, and 3:30 and 9:30 AM and PM, PDT during the summer. The weather SUMMARIES (discussed in Chapter 8) are updated at 5:00 AM, 11:00 AM, and 5:00 PM, PST, and 6:00 AM, 12:00 noon, and 6:00 PM, PDT. Both SUMMARIES and FORECASTS will be updated more often if potentially hazardous conditions arise.

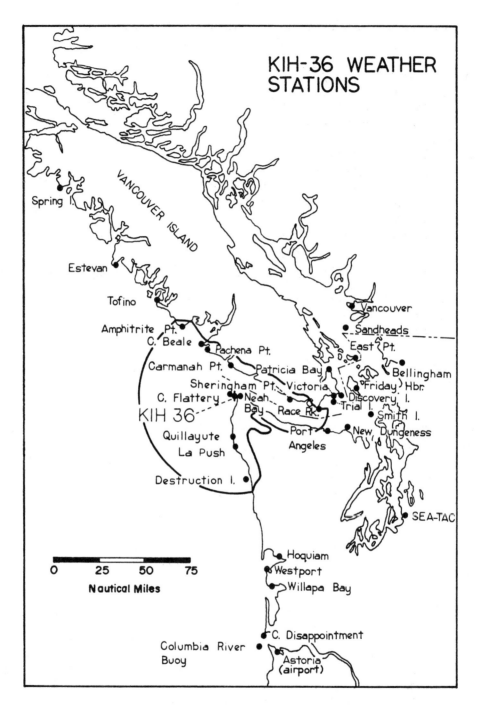

Figure 7-3 Weather observation points broadcast on KIH-36 VHF NOAA Weather Radio. Approximate area of reception shown by heavy line.

Weather Radio KIH-36

Frequency: 162.55 MHz ("Weather 1" or "WX1")
Transmitter location: Bahokus Peak near Neah Bay, Washington

Forecasts and weather observations are broadcast 24 hours a day. Contents are:

1. Station Identification
2. Coastal Zone Forecast, Coastal Zone Summary (general discussion of weather patterns, location of storms, etc.)
3. Marine Forecasts (updated as on KHB-60)
    a. Offshore (60 - 250 nm off Washington coast)
    b. Coastal (shore to out 60 nm)
    c. Strait of Juan de Fuca
    d. Inland Waters
    e. Canadian forecasts for Strait of Juan de Fuca and West Coast of Vancouver Island
    f. Grays Harbor Bar Forecast (updated 2:00 PM PST and 3:00 PM PDT)
4. Weather Observations (See Figure 7-3)
5. Coastal Forecast (for land areas along Washington coast)
6. Northwest Interior Forecast for Washington
7. Mountain Forecast
8. 6 - 10 Day Outlook
9. Special feature (if any), such as special discussion of holiday weather.

Comments about weather observation points and UPDATE TIMES for KIH-36 are the same as for KHB-60, listed earlier.

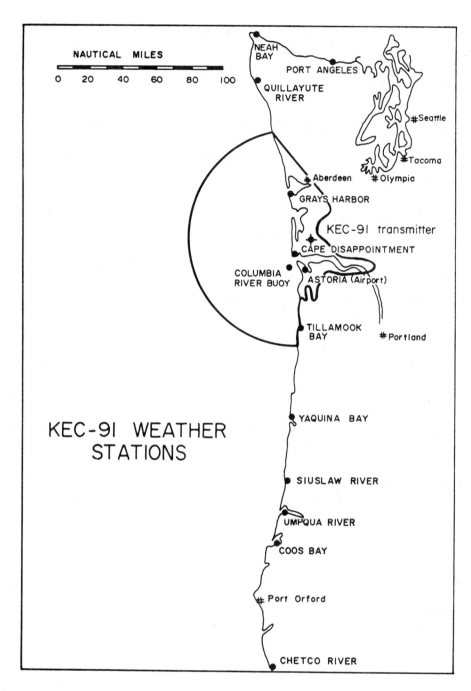

Figure 7-4 Weather observation points broadcast on KEC-91 VHF NOAA Weather Radio. Approximate area of reception shown by heavy line.

Weather Radio KEC-91

Frequency: 162.40 MHz ("Weather 2" or "WX2")
Transmitter location: Naselle Ridge about 15 miles NE of Ilwaco, Washington

Forecasts and weather observations are transmitted 24 hours a day. Contents of the transmission are:

1. Station identification.

2. State forecasts for western Washington and Oregon and forecast for Astoria and vicinity.

3. Marine Forecasts
    a. Coastal Forecast for Washington (updated 2:30 and 8:30 AM and PM PST and 3:30 and 9:30 AM and PM, PDT)
    b. Grays Harbor Bar Forecast (updated 2:00 PM PST and 3:00 PDT)
    c. Columbia River Bar Forecast (updated 10:00 AM and 10:00 PM PST and PDT)
    d. Coastal Forecast for Oregon (updated same as Coastal Forecast for Washington)
    e. Offshore Forecast (60-250 nm offshore) for area from C. Flattery, Wash. to C. St. George, Calif. (updated same as Coastal Forecasts).

4. Weather Observations. (Astoria and Columbia River buoy observations are updated hourly. Rest of the observations are only available from either 4:00 AM or 7:00 AM PST until 7:00 PM PST and are updated every 3 hrs.) See Figure 7-4.

5. Oregon Coastal Discussion (broadcast from 4:00 AM - 6:00 AM and 6:00 PM - 9:30 PM PST and PDT). This is a specially tailored weather discussion for mariners. A 48-hour outlook for marine weather is given.

6. Ocean Thermal Boundary bulletin (broadcast on a seasonal basis from April through October from 4:00 AM - 6:00 AM and 6:00 PM - 8:00 PM PDT). This broadcast contains information about the surface temperature structure of the ocean off Oregon and Washington. The information is updated once a week on Thursday beginning with the 6:00 PM broadcast.

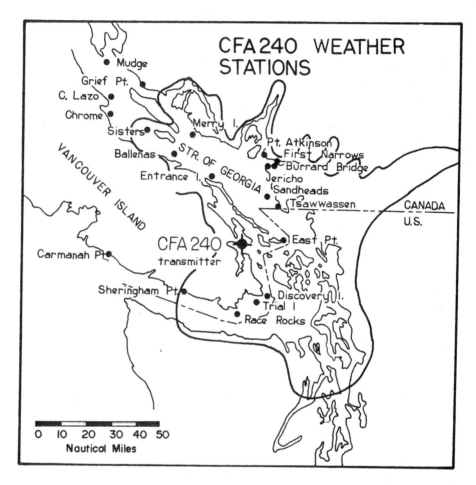

Figure 7-5 Weather observation points broadcast on CFA-240 Canadian Weatheradio. Approximate area of reception is shown by the heavy line.

Weatheradio CFA-240

Frequency: 162.40 MHz   ("Weather 2" or "WX2")
Transmitter location: Saltspring Island near Victoria, B.C.

Contents of the transmission are:

1. Weather synopsis for British Columbia
2. Area forecasts for:
    a. Greater Vancouver
    b. Lower Mainland
    c. East Vancouver Island and Sunshine Coast
    d. North and West Vancouver Island
    e. Greater Victoria
3. Western Washington public forecast
4. Cross Canada weather picture (carried approximately 7:00 AM to 11:30 PM).
5. Marine weather synopsis for British Columbia
6. Marine forecasts for:
    a. Georgia Strait
    b. Strait of Juan de Fuca
   (Forecasts updated at 5:00 AM, 11:00 AM and 7:00 PM PST)
7. Latest lighthouse reports (See Figure 7-5). Reports carried approximately 6:00 AM to 11:30 PM.
8. Mountain weather synopsis (winter months) with area forecasts for Vancouver Island mountains and South Coast mountains.

Weather Radio WXM-62

Frequency: 162.475 MHz   ("Weather 3" or "WX3")
Transmitter location: Boistfort Peak about 40 miles SSW of Olympia

Forecasts and weather observations are broadcast 24 hours a day. Contents of the transmission are:

1. Station Identification
2. Weather Synopsis for Southwest Washington and Forecast for Southwest Washington
3. Marine Forecast (C. Flattery to C. Disappointment out 60 nm and Inland Waters of Western Washington)
4. Grays Harbor Bar Forecast
5. Forecast for the Cascades and Olympics
6. Three to Five Day Outlook for western Washington
7. Weather Observations (sky condition and temperature only):
    Quillayute
    Hoquiam
    Astoria
    Portland
    Toledo
    Olympia   All observations updated hourly.
    Stampede Pass
    SEATAC Airport
    Bellingham
    Port Angeles

8. Mt. St. Helens Plume Trajectory Forecast

To date no studies have been completed that show the broadcast reception area, but initial indications are that reception is reliable to about 40 miles.

All VHF broadcasts are essentially line-of-sight transmissions, so the higher your antenna, the greater the distance of reception. Intervening land obstructions will weaken the signal or even block it though you may be within the reception areas shown in Figures 7-2 to 7-5. VHF weather radios may be incorporated as part of a multi-channel VHF marine radio or simply a small weather radio receiver you obtain at just about any department store radio and T.V. department. Regardless, these five stations we have just discussed are your surest bet for getting the most up-to-date information available.

Of course, the latest weather conditions are going to be more recent if you pick up on CB chit-chat or other conversations on VHF marine radios when people are talking about the weather where you want to go. But for reliability and the latest forecasts and warnings, the VHF weather radio network is your most useful source. During foggy conditions, conversation on VHF channel 14 (the Seattle Vessel Traffic Service channel) is extremely valuable in keeping you posted not only on the location of ship traffic but also on the degree of visibility along the Strait of Juan de Fuca and up and down the Sound.

Television and Radio

All major television stations in western Washington have sections of their news broadcasts dedicated to weather forecasts. One advantage with television is that you get a visual display of the weather as it exists and perhaps some satellite pictures which show you where the fronts lie.

Almost all radio stations in northwestern Washington broadcast forecasts during their regular programming. Those stations that also give the latest marine forecasts are listed in Table 7-1. Weather forecasts are made by NWS forecasters in Seattle and disseminated to the various stations.

TABLE 7-1 COMMERCIAL RADIO STATIONS WITH MARINE WEATHER BROADCASTS

Location	Station	Frequency
Aberdeen	KXRD	1320 kHz
Bellingham	KGMI	790 kHz
Bellevue	KZAM	1540 kHz
	KZAM-FM	92.5 MHz
Bremerton	KBRO	1490 kHz
Everett	KWYZ	1230 KHz
Mt. Vernon	KBRC	1430 kHz
Port Angeles	KAPY	1290 kHz
Raymond	KAPA	1340 kHz
Seattle	KAYO	1150 kHz
Seattle	KING	1090 kHz
	KING-FM	98.1 MHz
Seattle	KIRO	710 kHz
Seattle	KIXI	910 kHz
Seattle	KOMO	1000 kHz
Seattle	KOL	1300 kHz
Seattle	KVI	570 kHz
Tacoma	KTAC	850 kHz
Tacoma	KTNT	1400 kHz
	KTNT-FM	97.3 MHz
Vancouver	KISN	910 kHz

Telephone Recordings

The U.S. Coast Guard has four locations in Washington where you can obtain marine forecasts from telephone recordings. These are listed in Table 7-2.

TABLE 7-2 TELEPHONE RECORDINGS OF MARINE WEATHER

Location	Telephone Number	Update Schedule
Port Angles	(206) 457-6533	3 or 4 times daily
Neah Bay	(206) 645-2301	4 times daily
Grays Harbor	(206) 268-2622	4 times daily
Cape Disappointment	(206) 642-3565	4 times daily[a]

[a] or more often if bar conditions change

Newspapers

Most newspapers carry weather information and forecasts. However, because they are only published once or twice daily, the forecasts cannot be updated in fast changing situations, so their usefulness depends somewhat on the prevailing weather systems affecting our area. The weather maps included in some of the newspapers are useful because the types of fronts and general weather patterns are shown.

Radiotelephone Stations

The radiotelephone stations listed in Table 7-3 broadcast marine forecasts in the southern British Columbia and western Washington areas. The radio frequencies used are communications frequencies that are interrupted to provide weather broadcasts. The frequencies are on some marine VHF receivers and some marine single sideband (SSB) radios. The SSB frequencies can also be received on some portable shortwave radios. Also, some Radio Direction Finding (RDF) units (used primarily as a navigation aid) contain extra marine bands of frequencies that include radiotelephone frequencies.

High Seas Weather Transmissions

Detailed information about the weather pattern affecting the North Pacific Ocean from latitudes 30° N to 60° N, from the coast out to longitude 160° E, can be obtained from radiotelephone stations NMC and KMI located near San Francisco. The High Seas Bulletin originates at the National Weather Service Forecast Office located in Redwood City, California, and contains information such as the latitude and longitude of fronts, lows, highs, ridges, troughs, and strong wind areas; central pressures of highs and lows and forecasted movement; types of weather fronts and movement.

Station WWV in Fort Collins, Colorado and WWVH in Kekaha, Hawaii transmit less detailed information than NMC or KMI due to transmission time restrictions. WWV broadcasts storm information for the North Pacific east of 140° W, and WWVH transmits information about the North and South Pacific. WWV and WWVH are National Bureau of Standards time and frequency stations used mostly for time (GMT) and frequency standards for navigational purposes. These two stations are not radiotelephone frequencies.

High seas weather broadcasts in voice format from these four stations are summarized in Table 7-4.

TABLE 7-3 RADIOTELEPHONE STATIONS WITH MARINE WEATHER BROADCASTS[a]

Location	Call Sign	Frequency	Time (PST)[a]	Areas
Astoria, OR	NMW	157.1 MHz (CH 22A)	0915, 2115	C. Flattery to C. Disappointment out 60 mi
		2670 kHz	0403, 0933, 2133	C. Flattery to C. Lookout 60-250 mi offshore; C. Lookout to C. St. George 60-250 mi offshore; Warnings only for: C. Flattery - C. Disappointment out 60 mi., mouth of Columbia-Pt. St. George out 60 mi., Str. of J. de Fuca
Seattle[b]	NMW-43 (USCG)	157.1 MHz (CH 22A)	1030, 2230	Str. of Juan de Fuca, Inland Waters
Port Angeles	NOW (USCG)	2670 kHz	1015, 2215	Same as NMW Astoria
Camox, B.C.	VAC	2054 kHz[c] 161.65 MHz (CH 21)	0635, 0920, 1240, 1535, 2035	Str. of Juan de Fuca Str. of Georgia Johnstone Str. Queen Charlotte Str.
Victoria, B.C.	VAK	2054 kHz[c] 161.65 MHz (CH 21)	0450, 0615, 1210, 1450, 1920	W. Coast Vancouver Isl. Str. of Juan de Fuca Str. of Georgia
Vancouver, B.C.	VAI	2054 kHz[c] 161.65 MHz (CH 21)	0515, 0810, 1110, 1410, 1510, 2010	Str. of Juan de Fuca Str. of Georgia

[a] See "Worldwide Marine Weather Broadcasts" (Appendix J) for latest schedules.
[b] Transmitter on Mt. Constitution, Orcas Island.
[c] Single sideband, full carrier

TABLE 7-4 HIGH SEAS WEATHER BROADCASTS

Call Sign	Frequency[a] (kHz)	Time[b] (PST)	Call Sign	Frequency[a] (kHz)	Time[b] (PST)
NMC	4428.7 8765.4 13113.2 17307.3	0230 & 2030	KMI	4357.4 8728.2 8784.0 13100.8 13187.6 17236.0	0700, 1600, 2200
	8765.4 13113.2	0830 & 1430		4403.9 8743.7 13103.9 13107.0 17239.1 22636.3	0500, 1100
WWV	2500.0 5000.0 10000.0 15000.0 20000.0	8 min. past each hour.	WWVH	2500.0 5000.0 10000.0 15000.0 20000.0	48 min. past each hour

[a] Amplitude modulation is single sideband, suppressed carrier.
[b] See "Worldwide Marine Weather Broadcasts" (Appendix J) for latest schedules.

FAA Radio Transmissions

There are four aircraft electronic navigation transmitting sites operated by the Federal Aviation Administration in northwestern Washington that also carry weather observations, weather synopses, and aviation forecasts. The stations are referred to as VOR stations (Very High Frequency Omni-Directional Range). Although intended for aviation use, the information can be useful to the boater having a radio capable of picking up the transmissions. The frequencies are just above the commercial FM band but below the marine VHF-FM band

One advantage of these weather observations, which are only from airports, is that they are updated every hour; the disadvantage being that they are usually not too representative of over-the-water conditions. You can, however, use the pressure values to keep track of changes in the pressure and note changes in the pressure gradient. Information in these broadcasts can sometimes help you judge whether or not a frontal system -- predicted in the less frequent marine broadcasts -- is arriving on time. The VOR sites in our area with weather broadcasts are listed in Table 7-5.

Aviation weather that may be of similar value is also broadcast continuously on AERO beacon transmissions. The frequencies are in the same band as marine RDF beacon frequencies, just below commercial AM frequencies. AERO beacons are RDF beacons intended for aircraft navigation but also occasionally of value to marine navigation. Some AERO beacon frequencies are listed on nautical charts. They can be received on marine RDF radios. A useful frequency for western Washington is listed in Table 7-5. Racing sailors may find the hourly updates on this station a valuable aid to predicting the times that frontal systems are expected to pass over Puget Sound. RDF radios are required equipment on several classes of racing boats.

TABLE 7-5 VOR AND AERO SITES WITH CONTINUOUS WEATHER BROADCASTS

Location	Frequency	
Tatoosh	112.2 MHz	
Paine Field, Everett	114.2 MHz	VOR STATIONS
SEATAC Airport	116.8 MHz	
Olympia	113.4 MHz	
Seattle (BF)	362 kHz	AERO STATION

[a] Morse code identifiers

Vessel Traffic Service

The Puget Sound Vessel Traffic Service (PSVTS) is the U. S. Coast Guard division that monitors and provides advice to ship traffic in Puget Sound and the Strait of Juan de Fuca. PSVTS also has the authority to control vessel movements when necessary during conditions of vessel congestion, adverse weather, reduced visibility, or other hazardous conditions -- though the shift from advice to authority is rarely required.

The Vessel Traffic Service is not an official source of weather information, but in restricted visibility they can be an extremely valuable practical source of up to date visibility conditions in the Strait of Juan de Fuca and Puget Sound. They coordinate ship traffic using VHF channel 14. In restricted visibility it is highly recommended that all vessels equipped with VHF radio passively monitor channel 14. You will not only learn of nearby traffic, but you will also keep abreast of visibility conditions as these are discussed frequently between participating traffic and the VTS center.

Chapter 8
PUTTING IT ALL TOGETHER

Since VHF NOAA Weather Radio and VHF Weatheradio Canada are the most up to date sources of weather information easily accessible to us, we should put it to its intended use -- helping us make reasonable decisions. Even if all we paid attention to were the forecasts, we would be right about the weather more times than not, but we can extend our knowledge a lot further if we make use of the weather synopsis (called "weather summary" on VHF) and the actual weather observations. It was mentioned earlier that weather observations are only updated every three hours on VHF, which makes their usefulness considerably less two hours later. However, we can still use them to detect trends. These actual weather observations, when used with the forecast and weather synopsis, will give us a good idea about the entire weather situation, especially if we use all this information along with our knowledge about the principles of the weather.

Using the Weather Summary

The weather summary gives a general description of the large scale weather patterns affecting our area and tells us such things as the location of highs, lows, and fronts and how they are moving and changing. The synopsis is not a forecast as such, but part of the weather forecast is based on the synopsis because the forecaster has to have a starting point in the weather pattern to begin formulating his predictions. If the synopsis says a low or depression is deep or intense, this means that gale (34 - 47 kn.) to storm force (over 47 kn.) winds are present around the low over the open ocean. Depending on the movement of the low and any fronts associated with it, we may or may not get these strong winds in the Inland Waters, but we are considerably more likely to get them over the Strait of Juan de Fuca if the storm system gets close enough.

Troughs

The synopsis might also warn of potentially high winds by informing us that a "trough" of low pressure is off the coast. Weather associated with a trough is often very cloudy and rainy (or snowy) with blustery winds. A trough is also an area where fronts may form, which will make the weather even nastier. A trough is a special part of a large low pressure system in which the isobars are stretched out in the form of "V"s along a line connecting the points of the

"V"s with the center of the low. The sharper the points on the "Vs" the worse the winds and weather will be.

Ridges

Terms such as deepening or intensifying used to describe a low in the weather summary mean that the low is showing a decrease in pressure; hence, an increase in the winds around the system is occurring. The opposite term, filling, means that the central pressure in the low is increasing, which is usually followed by decreasing winds.

When we hear that a ridge is building into western Washington, our barometer should indicate a rise in pressure, and the weather should improve. Depending on the time of year, we could get some fog. A ridge is to the high pressure system what the trough is to the low pressure system, but instead of having pressure decrease as you go from the outer "V" toward the center, the pressures increase. It's kind of like climbing up the crown of a ridge toward a mountain top in which the peak is the center of the high with the highest pressure. Carrying the analogy to the air, we find that as the air flows around the high (like an airplane circling the mountain peak) it will climb up one side of the ridge and drop down the other side as it continues its clockwise journey (Chapter 2).

In the process of climbing the ridge, clouds may form and precipitation fall because the air is cooling by expansion, causing the water vapor to condense. Conversely, as the air flows down the other side of the ridge, it heats by compression, which will likely cause the clouds to evaporate. The clouds may not always vanish with a ridge, however, since surface heating can cause cumulus clouds to form, and there is the possibility of low level stratus clouds coming in from the ocean. In general, we can expect improving weather as the ridge approaches us and passes directly over us, followed by increasing clouds on the west side of the ridge. This general rule depends on the orientation of the ridge and does not always apply.

Factors making up the weather are complex to say the least, but using the weather summary and knowing in a general way what types of weather are associated with various weather systems, we can reduce our chances of being caught off guard if the forecast goes sour or misses the trend. For example, if we are under the influence of a high, and the weather summary says that the high is weakening (barometric pressure decreasing) because a low is approaching the coast along with a weather front, then the forecast will call for increasing southerly winds and cloudiness for the Inland Waters and increasing easterly winds and cloudiness for the Strait of Juan de Fuca. We can expect the present weather to change. If it does not, then we know that the high did not weaken as expected or is weakening much slower. Let's take a look at how we can put our weather knowledge to use by examining an actual weather situation and the weather summaries, forecasts, and observations that went with it.

Example of Using Forecasts, Summaries, and Observations

It is a Tuesday evening during a fine weather period in November, the day before our outing. After looking at the newspaper weather map and watching a television weather show during the news broadcast, we learn that a front (occluded type from the weather map) is expected to move across western Washington as a high pressure system over us weakens and moves eastward. We plan to take our boat from the Des Moines Boat Harbor up to Edmonds for an all-day outing, returning to Des Moines at around 8:00 p.m. Wednesday.

The next day we tune in KHB-60 NOAA Weather Radio shortly after 8:30 a.m. to catch the latest weather summary, Inland Waters forecast, and weather observations for the Puget Sound area. We also will keep track of the pressures at SEATAC and Bellingham to see how the pressures change throughout the day. All this information will give us a good idea about the trend in the weather and how fast it is changing, which, in turn, will help confirm the forecast.

Referring to the Broadcast and Update Times listed with the KHB-60 description in Chapter 7, at 8:30 a.m., then, we obtain the weather summary for 5:00 a.m. (summaries are made at 5:00 a.m., 11:00 a.m., and 5:00 p.m. PST); the latest Inland Waters forecast issued at 8:30 a.m.; and the 7:00 a.m. weather observations and pressures. (Remember, weather observations are only updated every 3 hours.) Since it is hard to record all this information, we should just keep the salient points of the weather summary in our head, namely, the movement and relative strength of highs, lows, and weather fronts. Similarly, recording the entire forecast is difficult, so we might only record the important points about the wind. The weather observations can be recorded on a piece of paper we have prepared beforehand listing the stations of interest (Figure 7-2). A vertical time sequence for every 3 hours is useful because we can spot trends at a glance. The complete information as broadcasted is shown in Table 8-1

5 AM WEDNESDAY WEATHER SUMMARY:

AN INTENSE LOW PRESSURE SYSTEM IS LOCATED IN THE EASTERN PACIFIC. TWO FRONTAL BANDS ARE ASSOCIATED WITH THIS SYSTEM. SATELLITE PICTURES SHOW THAT THE FIRST FRONT HAS BEEN MASKED BY THE SECOND MUCH STRONGER FRONT. THIS FRONT IS MOVING NORTHEASTWARD AT ABOUT 25 MPH WHICH PLACES THE BACK EDGE OF THE CLOUD BAND ON THE COAST BY THURSDAY AFTERNOON. MOST OF THE MOVEMENT OF THIS SYSTEM IS NORTHEASTWARD SO THE WEATHER WILL DETERIORATE SLOWLY LATE TODAY AND TONIGHT. GUSTY WINDS WILL ACCOMPANY THE CLOUDS AND RAIN.

0830 AM WEDNESDAY FORECAST FOR INLAND WATERS OF WESTERN WASHINGTON:

WINDS SOUTHEASTERLY 10 TO 15 KNOTS INCREASING TO 15 TO 30 KNOTS MAINLY NORTH PART TONIGHT BECOMING SOUTH TO SOUTHWEST 10 TO 20 KNOTS THURSDAY. MOSTLY FAIR TODAY. INCREASING CLOUDS AND RAIN LIKELY TONIGHT. SHOWERS THURSDAY.

From the weather summary and forecast, we see that conditions are expected to worsen, especially in the northern part of the Inland Waters by tonight. Winds are expected to increase from the southerly direction during the day, so we should keep in mind that we'll be bucking head seas all the way back from Edmonds. The 7:00 a.m. weather observations indicate that winds are 10 to 17 knots from the south at Alki and West Point, which appears to be contrary to the pressure gradient between Bellingham and SEATAC. Lower pressure at SEATAC indicates that the wind should be from the north instead of southerly direc-

TABLE 8-1 KHB-60 SELECTED WEATHER OBSERVATIONS FOR WEDNESDAY

Pressure (millibars)

	SEATAC	BELLINGHAM	DIFF.	SEATAC	OLYMPIA	TACOMA	ALKI PT.	WEST PT.	PT. NO PT.	BURROWS I.	WHIDBEY
7 AM	1026	1027	-1 mb	ESE 10 Fair	SE 4 Foggy	Calm Foggy	S 10	S 17 Clear	Missing	NW 2 G13	SE 12 G21 Partly Cldy
10 AM	1025	1026	-1 mb	ESE 10 Fair	Calm Partly Cloudy	Calm Fair	S 7	S 12 Partly Cloudy 1' waves	S 10 Partly Cloudy 1' waves	E 4 G13	SE 15 G22 Fair
1 PM	1024	1023	+1 mb	SE 12 Fair	E 5 Cloudy	Calm Partly Cloudy	S 6	S 8 Partly Cloudy	Calm Partly Cloudy	NE 5 G14	SE 16 G20 Cloudy
4 PM	1022	1021	+1 mb	ESE 13 Cloudy	WSW 4 Cloudy	Calm Cloudy	S 6	S 11 Partly Cloudy	Missing	E 5 G17	SE 16 G18 Cloudy
7 PM	1020	1019	+1 mb	ESE 13 Cloudy	SE 4 Cloudy	Calm Cloudy	S 4	S 14 Partly Cloudy	Missing	SW 9 G25	SE 18 G26 Cloudy
10 PM	1019	1017	+2 mb	SE 10 Cloudy	Calm Cloudy	Calm Cloudy	S 10	Missing	Missing	ESE 12 G23	SE 17 G25 Cloudy
1 AM Thur.	1019	1017	+2 mb	E 9 G23 Rain	W 3 Rain	E 6 Rain	S 7	Missing	Missing	S12 G30	SE 24 G29 Rain

Note: Wind speeds are in knots. Wind directions are the TRUE directions from which the winds are blowing (not magnetic directions). A "G" indicates a gusty wind. For example, Burrows I. at 7:00 a.m. has a steady NW wind at 2 knots with gusts to 13 knots.

tions in the Puget Sound area. However, the difference in pressure between the two stations is only 1 millibar. When pressure differences between these two stations are 1 mb or less, the winds may not conform to the pressure gradient principal (Chapter 2).

We have already examined the tidal current tables and determined that steepening of the waves is not too likely since the currents are less than one knot. Referring to the Wave Height Tables in the Appendix, we can get an idea of the height of waves already in the area we'll be traveling through. Location 29 (Puget Sound west of Edmonds) and Location 30 (Puget Sound west of West Point) will be the points to use in the Wave Tables. Averaging the 10 and 17 knot wind speeds, we get 13.5 knots (round to 14 knots). We assume that the wind has been blowing this strong since 7:00 a.m., the time of observation. Let us assume it is now 9:00 a.m. The wind has been blowing 14 knots for at least a two hour period, which will create 1.3 foot waves at both points as determined from the Wave Tables. We had to interpolate wave values between 10 and 20 knots using the 14 knot speed. This wave height isn't too bad while we are going with the waves, but will be quite choppy on our way back from Edmonds if the winds don't drop. From the forecast and the weather summary we really can't expect them to decrease because a front is off our coast; yet, on the other hand, the forecast doesn't predict a large increase in the wind speed for the southern part of the Inland Waters. We can get some idea of how fast the weather pattern is changing by looking at the drop in pressure at Bellingham and Seattle throughout the day. The slower the change, the slower the change, the slower the pattern is changing.

As long as we're prepared for a lot of spray and a longer trip when we turn around to come back, the weather shouldn't be too bad. However, we had better make sure we have plenty of fuel since we will be encountering head seas the whole way back.

We set out on our journey under clear skies and a chop on the water, but it's not uncomfortable because we are heading with the wind and seas, which also reduces the wind chill factor. Shortly after 11:00 a.m., we listen to KHB-60 again to get the latest weather summary and observations. The forecast will be the same one we got at 8:30 unless conditions have changed enough to warrant an update.

11 AM WEDNESDAY WEATHER SUMMARY:

HIGH PRESSURE OVER THE PACIFIC NORTHWEST CONTINUED TO HOLD OFF PACIFIC WEATHER DISTURBANCES WEDNESDAY MORNING. SATELLITE PICTURES THIS FORENOON SHOW A WIDE FRONTAL CLOUD BAND JUST OFFSHORE AND SLOWING IN ITS EASTWARD SWING INTO THE AREA OF HIGH PRESSURE. THIS WEATHER PATTERN WILL LEAD TO INCREASED CLOUDINESS IN WESTERN WASHINGTON TODAY WITH RAIN TONIGHT ALONG THE COAST THEN SLOWLY INVADING THE PUGET SOUND BASIN THURSDAY.

The latest summary shows that the weather pattern has not changed much since we got the 5:00 a.m. summary. The weather observations show a slight decrease in the winds, but we should note that the pressure has dropped 1 mb at both Bellingham and SEATAC, although the pressure gradient has remained the same between them. We also notice that it's getting cloudier. We continue on, keeping in mind that the front must be getting closer. The seas continue to be choppy with the southerly wind.

Around 1:30 p.m., we listen to KHB-60 again to get the latest observations that were taken at 1:00 p.m. and are surprised when a new forecast is on the air instead of the regular time at 2:30 p.m. The forecaster has detected a more rapidly changing situation, which required an updated forecast along with a Small Craft Advisory.

1 PM WEDNESDAY FORECAST FOR INLAND WATERS OF WESTERN WASHINGTON:

. . . SMALL CRAFT ADVISORY. INCREASING SOUTHEASTERLY WINDS TO 15 TO 30 KNOTS WITH GUSTS OF 45 KNOTS NORTH OF EVERETT THROUGH TONIGHT. WINDS DECREASING THURSDAY TO SOUTHERLY 15 TO 25 KNOTS AND CONTINUING THURSDAY NIGHT. PERIODS OF RAIN THROUGH THURSDAY NIGHT.

It looks like the winds are expected to become rather strong over the Inland Waters north of Everett by tonight and to be up to 30 knots from the southeast in lower Puget Sound. So far, from the observations we recorded for 7:00 a.m., 10:00 a.m., and 1:00 p.m., we see that the winds have shown no increase to speak of in lower Puget Sound, but are getting stronger north of Everett at Burrows Island and Whidbey.

The pressure at SEATAC has dropped from 1026 mb to 1024 mb and at Bellingham has dropped from 1027 mb to 1023 mb where it is now one millibar lower than at SEATAC. This is what we expect with a front approaching the coast from the west. We are now faced with a judgement call: should we turn around and head back to Des Moines or should we continue as planned and get back at 8:00 p.m., some three hours after daylight has ended?

With Small Craft Advisories in effect, the fact a front is on the way, and because winds are already fairly strong in northern Puget Sound, the prudent thing to do would be to turn around now and get back to Des Moines before dark. If the winds do reach 30 knots from a southerly direction for even one hour, we can see from the Wave Tables that seas will be 3.2 feet off West Point, which makes for an extremely uncomfortable and wet ride in our boat.

We return to Des Moines by 5:00 p.m. and are on our way home by 5:30. Wondering how things turned out, we listen to the radio again and get the 5:00 p.m. weather summary and the 4:00 p.m. weather observations. It really turned out to be one of those "iffy" situations in which the weather only slowly deteriorated in the southern Inland Waters, but did grow much worse in the northern sections, as can be seen using the observations from Burrows Island and Whidbey. The pressure gradient doubled between SEATAC and Bellingham, and the pressures continued to drop at both stations. We exercised good judgement in this case because it could easily have gotten much worse in the southern areas, too.

5 PM WEDNESDAY WEATHER SUMMARY:

HIGH PRESSURE OVER THE PACIFIC NORTHWEST CONTINUED TO HOLD OFF PACIFIC WEATHER DISTURBANCES WEDNESDAY MORNING. SATELLITE PICTURES STILL SHOW A WIDE FRONTAL BAND JUST OFFSHORE AND SLOWING IN ITS EASTWARD SWING TOWARD WASHINGTON BECAUSE OF THE HIGH PRESSURE. THIS WEATHER PATTERN WILL LEAD TO INCREASED CLOUDINESS

IN WESTERN WASHINGTON TODAY AND HIGH WINDS IN COASTAL AREAS WEDNESDAY EVENING. RAIN IS EXPECTED TO BEGIN ON THE COAST TONIGHT THEN SLOWLY INVADE THE PUGET SOUND BASIN THURSDAY.

The weather observations after 4:00 p.m. are included to show what finally happened. As can be seen, weather conditions gradually worsened into the night. When using actual weather observations, be sure to bracket the area you plan to operate in so that a broader picture of the weather situation can be obtained. In the example above, we would not have learned much if all we used was the observation at West Point, for example. Sometimes, a particular observation may be omitted because of communication or instrument problems, which would leave us with no information if all we were relying on was that one point. We could have learned even more by also using the observations from the Strait of Juan de Fuca as well as the pressure at Quillayute. This additional information would be useful in tracking a weather front through western Washington. KHB-60 Weather Radio broadcasts pressures at SEATAC, Bellingham, Quillayute, Astoria, Oregon, and North Bend, Oregon. KIH-36 Weather Radio does not broadcast pressures except for SEATAC.

It goes without saying that recording all this information in a small boat that is bouncing all over the place is not an easy task, so we are limited to the number of observations we can keep track of. We should at least record one of the stations giving pressure and no less than two stations bracketing our boating area. Finally, we should also estimate the wind velocity at our location and watch the sky.

Weather Check List

Trying to guess what the weather and waves are going to do next is a challenging and often times fascinating game in which we have to expect that we will be wrong some of the time. We have covered many aspects of the weather in this book, but the basic concepts can all be summarized in a weather check list.

<u>WEATHER CHECK LIST</u>

1. If you are near the area you plan to operate in, what is the state of the present weather?
2. Use marine forecast to determine the trend (better, same, worse?). Winds?
3. Marine warnings. Small Craft Advisories: be cautious. Gale Warning: be very careful and don't get caught in unprotected waters if you or your boat can't handle rough seas. Storm Warning: stay ashore.
4. Weather summary. Location of fronts? Slow or fast changing weather?
5. Latest weather reports bracketing area where you plan to operate.
6. Use Wave Tables to estimate seas.
7. Tidal currents. With or against the wind and waves?

The final question to answer: Are the risks acceptable to you?

APPENDICIES

APPENDIX CONTENTS

page

Appendix A	Glossary	123
Appendix B	Wind and Wave Descriptions	126
	Beaufort Wind Scale	127
Appendix C	Wave Height Tables	
	Inland Waters and Strait of Juan de Fuca	128
Appendix D	Wave Height Tables	
	Open Ocean	138
Appendix E	Water Temperatures in Northwest Washington	141
Appendix F	Hypothermia	142
Appendix G	Wind Chill	144
Appendix H	Conversion Tables	
	Pressure: Inches to millibars	145
	Temperature: °F to °C	146
Appendix I	Cloud Identification	147
Appendix J	Additional Reading	150

APPENDIX A

GLOSSARY OF COMMONLY USED TERMS IN WEATHER FORECASTS

BACKING -- Counterclockwise change in wind direction (e.g., change in direction from north to northwest to west; east to northeast to north, etc.).

COMPLEX LOW -- A large area often more than 1000 nm across in which two or more low centers exist.

DEEP LOW -- A rather subjective term used to describe the central pressure of a low center (usually when it is about 975 mb or less). Winds are strong gale force to storm force around the low.

DEEPENING LOW -- A low in which the central pressure is decreasing with time. Winds would be expected to increase as the low deepens.

DENSE FOG -- Fog in which visibility is less than 1/4 mile.

DEVELOPING HIGH -- A change in the weather pattern in which higher pressure is building up over an area.

DEVELOPING LOW -- A change in the weather pattern in which lower pressure is forming over an area that is likely to result in a definite low center.

DISSIPATING LOW -- A low center that is becoming weaker as the central pressure increases with time. Winds, in most cases, decrease; low expected to vanish.

EASTERLY WINDS -- True wind direction from the northeast to southeast sector; used when the forecaster is uncertain about the exact wind direction, but is confident that it will come somewhere between NE and SE.

FEW SHOWERS -- Low probability of precipitation in which a small number of showers will occur.

FILLING LOW -- Low center in which the central pressure is increasing with time; not the same as dissipating low because the low may not necessarily vanish.

FRONT -- Boundary zone separating two masses of air, one of which is colder than the other. Types of fronts: (1) warm front (warmer air overtaking colder air); (2) cold front (colder air overtaking warmer air); (3) occluded front (cold front overtaking warm front); (4) arctic front (special case of cold front in which air behind front is very cold, say less than 10°F); (5) stationary front (a front that is not moving). See Table 1-1 and 5-1 for winds and weather associated with moving fronts.

GALE -- Sustained wind speed of 34 through 47 knots.

GALE WARNING -- Special alert to mariners for sustained winds of 34 through 47 knots.

HIGH -- An anticyclone. A weather pattern in which isobars at the center of the pattern are of higher pressure than those further from the center. Winds are clockwise in the Northern Hemisphere.

INTENSIFYING HIGH -- Anticyclone in which the central pressure is increasing.

LOCALLY STRONGER WINDS -- Conditions in which winds over many small areas too numerous to mention are expected to be higher than the general wind in the area covered by the forecast (e.g., locally stronger winds may occur in fjords and channels as compared to winds over the open waters).

LOW -- Cyclone or depression. Weather pattern in which closed isobars at the center of the pattern are of lower pressure than those further from the center.

MODERATE LOW -- A rather subjective term used to describe the intensity of a low; used when the central pressure is about 975 to 1000 mb. Winds generally less than about 40 knots.

NORTHERLY WIND -- True wind direction from the NE to NW sector.

NUMEROUS SHOWERS -- Frequent number of showers are likely over more than half the area covered by the forecast.

PATCHY FOG -- Fog occurring in less than half the area covered by the forecast.

PERIOD -- Time (in seconds) it takes for successive wave crests (or troughs) to pass a fixed point.

PRESSURE GRADIENT -- Difference in pressure between two points divided by the distance between them. The greater the difference in pressure between the same two points, the greater the wind. The closer the isobars are together on a weather map, the greater the pressure gradient.

RIDGE -- Area of high pressure in which the isobars are elongated instead of circular or nearly circular. Flat ridge: elongation of isobars is not great enough to prevent penetration of weather fronts through the ridge. Strong ridge: elongation of isobars is great enough to prevent weather fronts from penetrating the ridge. The concept of a ridge also applies to upper air weather maps that are made for heights over 5,000 feet.

SCATTERED (SCT) SHOWERS -- Precipitation is expected to fall within about 30% to 45% of the area covered by the forecast.

SEAS -- Combination of wind waves and swell making up the irregular surface of the sea. Mariners often use the term to mean wind waves only, which is not the same meaning used by forecasters. Forecasters will usually use the term "combined seas" instead of "seas" to avoid this ambiguity.

SHORT WAVE -- Term occasionally used in forecasts that may be considered to be a moving low, weather front, or high.

SIGNIFICANT WAVE HEIGHT -- Average of the highest one third of the waves.

SOUTHERLY WIND -- True wind direction from the SE to SW sector.

STATIONARY (STNRY) -- Less than five knot movement of high, low, or front.

STORM -- Low pressure system in which winds are 48 knots or higher.

STORM WARNING -- Highest level of marine warning for storms that are not hurricanes or typhoons. Issued when winds are expected to be 48 knots or higher.

STRONG LOW or STRENGTHENING LOW -- Low pressure system in which the winds are at least gale force or are likely to reach gale force or higher as the central pressure of the low drops.

SWELL -- Waves that have left the wind fetch area where they were created and have become more rounded in shape and regular in period.

TROUGH -- An elongation of the isobars around a low. Inclement weather often occurs in a trough. This term is also applied to weather patterns found in the upper air.

VEERING -- A clockwise change in wind direction, as from west to northwest.

WEAK LOW -- Low pressure system with winds less than gale force. Central pressure is usually above 1000 mb.

WEAK FRONT -- Weather front that has winds less than about 20 knots and which is disappearing with time. A strong front may become a weak one as it tries to push through a stationary ridge of high pressure. Some cloudiness and light precipitation may still occur.

WEAK HIGH -- High pressure system that is incapable of keeping weather fronts from passing through it. Central pressure of a weak high will usually be 1020 mb or less.

WESTERLY -- True wind direction from the NW to SW sector.

WIND WAVES -- Waves caused by the wind in a given area.

APPENDIX B

WIND AND WAVE DESCRIPTIONS

Wave descriptions are used to categorize a certain appearance of the sea. They do not describe how comfortable your ride will be since this depends on the size of your craft and the direction you are going in relation to the waves.

These descriptions also do not take into account the effects of current on wave steepness. For example, a 12-kn wind blowing with a 1 or 2-kn current may not raise breaking crests on wavelets ("popcorn"), whereas the same wind blowing against this current will show occasional white crests. In this wind range, this can often help you spot current boundaries on the water surface.

If you want to measure your own wind speed, you can obtain very low-priced, handheld anemometers from just about any well-stocked yachting supply business. One particular brand, the Dwyer Anemometer, is very reasonably priced and will fit in a coat pocket.

The National Weather Service does not use wave descriptions for wave-height reports in the Inland Waters or the Strait of Juan de Fuca. If heights are given on the radio, they will be in feet.

TABLE B-1 WAVE DESCRIPTIONS BY APPEARANCE

Wind Speed (knots)	Wave Appearance and Effects
1 - 4	Ripples, no foam. Flags limp.
5 - 9	Small wavelets, no foam. Wind felt on face; small flags wave.
10 - 15	Large wavelets, crests begin to break. Light flags extended.
16 - 27	Moderate waves, many white caps, some spray. Wind whistles in rigging.
28 - 40	Sea heaps up with spindrift and foam streaks. Walking resistance high.
41 - 48	Dense streaks of foam, much spray. Loose gear and light canvas may part.
49 - 55	Very high waves with breaking crests. Visibility reduced.
55 - 65	Very high breaking waves. Very poor visibility due to flying spray.

TABLE B-2   WAVE DESCRIPTIONS USED BY CANADIAN ATMOSPHERIC ENVIROMENT SERVICE

Description	Meaning
Smooth	glassy or occasional ripples
Rippled	wavelets with crests 1/2 ft. or less above troughs
Choppy	wave crests 1 - 2 ft. above troughs
Moderate	wave crests 2 - 6 ft. above troughs
Rough	wave crests over 6 ft. above troughs

TABLE B-3   BEAUFORT WIND SCALE[a]

Beaufort Force Number	Wind Range (knots)	Description[b]
0	0 - 1	Calm
1	1 - 3	Light Air
2	4 - 6	Light Breeze
3	7 - 10	Gentle Breeze
4	11 - 16	Moderate Breeze
5	17 - 21	Fresh Breeze
6	22 - 27	Strong Breeze
7	28 - 33	Near Gale
8	34 - 40	Gale
9	41 - 47	Strong Gale
10	48 - 55	Storm
11	56 - 63	Violent Storm
12	64 - up	Hurricane

[a] Sea state descriptions and photographs given in: Bowditch's _American Practical Navigator, Vol. 1_

[b] According to the World Meteorological Organization (1964)

APPENDIX C

WAVE HEIGHT TABLES: INLAND WATERS AND STRAIT OF JUAN DE FUCA

The tabulated wave heights are in feet and are based on the speed and duration of the wind. The wind directions are TRUE directions, not magnetic. If the wind direction is within one compass point (e.g., SSE to SSW for a S wind shown in the table), the values of heights will give a reasonable estimate of the significant wave height (the average height of the highest one third of the waves; see Chapter 2). This is the height very close to the one which would be reported by an experienced seaman. The occasional _highest_ waves may be up to 90% higher than the tabular values and are the ones that give us the most problems because they are so much larger. Values shown in the tables are for areas having currents less than 3 knots and no ocean swell. If a current runs against the waves, they become much steeper and higher; if the current goes with the waves, they will be lower. Using the tables is easy:

1. Pick point of interest.
2. Select stations which bracket the area of interest and make your own estimate of the true wind speed and direction if you are in an area exposed to the wind. (See Figures 7-2 and 7-3 for wind observation points reported on VHF Weather Radio).
3. Use forecasts and observations to determine duration of wind in hours.
4. Read wave height based on direction, speed, and duration of wind. You may have to interpolate between values.

You will note that many of the tables do not have wave heights for winds lasting beyond a certain number of hours. This is because the maximum wave development for a given wind speed takes place in a given time period so that no matter how much longer it blows at that speed, the waves won't get any higher. They will get higher, however, if the tide changes and a strong current runs against the waves.

This example shows how you can make an estimate of the expected wave height for a mid-channel area. Be sure to select as many observation points as possible that bracket the area you are interested in.

Let's assume we have listened to the VHF Weather Radio at 6:30 a.m. PST, catching the weather observations for 4:00 a.m. (remember, they are only updated every 3 hours starting at 1:00 a.m.). We plan to depart Port Angeles at 7:00 a.m. for Victoria B.C. The forecast for the Strait of Juan de Fuca predicts continuation of strong east to southeast winds to 30 knots through the morning. Record the wind observations. Figure C-1 shows the wind observations plotted on a map of the area so it is easier to see what is happening. Refer to the Wave Height Tables for "Location 3 with ESE wind," as this is the point closest to the area of concern. From the 4:00 a.m. observations and the forecasts, we can conclude that winds have been blowing at least 3 hours at about 25 knots (4 to 7:00 a.m.). Note that the winds range from 20 to 30 knots across the area, except for Port Angeles, which is really not representative because it is sheltered from the E to SE winds. Using a wind duration of 3 hours, we interpolate between wave values for 20 and 30 knots (3.2 plus 5.7 divided by 2) to arrive at 4.5 feet. This is the average height of the highest one-third of the waves.

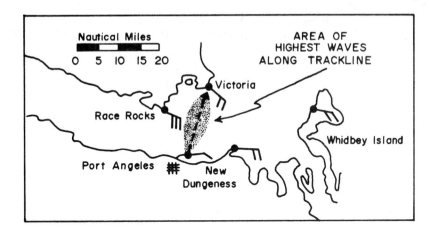

Figure C-1 Plot of wind observations in vicinity of Port Angeles.

Example of wave-height prediction table

Location 3:   Strait of Juan de Fuca,
              mid-channel, N. of Port Angeles

Location 3 with ESE Wind

WIND (kn)	WIND DURATION (hr)						
	1	2	3	4	5	6	7
10	--	0.3	0.8	1.2	1.3	1.4	1.5
20	--	1.8	2.7	3.2	3.6	3.7	
30	--	3.2	4.6	5.7	6.3		
40	--	4.8	7.0	8.6			
50	--	6.7	9.8	11			
60	--	8.6	13	14			
70	--	11	16	17			

AREAS WITH PREDICTED WAVE HEIGHTS IN TABLE C-1

LOCATION NUMBER ON FIG. C-2		WINDS USED FOR WAVE PREDICTIONS		
1	Cape Flattery	ESE		
2	Strait of Juan de Fuca (mid-channel N of Clallam Bay)	ESE	WNW	
3	Strait of Juan de Fuca (mid-channel N of Port Angeles)	ESE	WNW	NE
4	Smith Island (2.9 nm SE of)	N	W	NW
5	Admiralty Inlet (vicinity Point Wilson)	N		
6	Admiralty Head (0.5 nm S of)	SE	WNW	
7	Strait of Georgia (9 nm SE of Point Roberts)	SE	WNW	NE
8	Haro Strait (2.9 nm of Pyle Point, San Juan I.)	SE		
9	Prevost Passage (1.6 nm W of Turn Point, Stuart I.)	SSE	NE	
10	Spieden Channel (east end of)	NE	SE	WNW
11	San Juan Channel (1.2 nm W of Neck Point, Shaw I.)	SE	NW	
12	San Juan Channel (2.6 nm S of Shaw I.)	SSE		
13	Cherry Point (1 nm SE of)	S		
14	Rosario Strait (1.6 nm NW of Reef Point, Cypress I.)	N	S	
15	Bellingham Bay (3 nm SE of Fish Point)	SSE		
16	Samish Bay (2 nm NE of William Point)	NNW		
17	Padilla Bay (3.5 nm SE of William Point)	NW		
18	Saratoga Passage (2 nm W of Rocky Point, Camano I.)	SSE		
19	Saratoga Passage (2 nm W of Camano Head, Camano I.)	NE	SE	
20	Port Susan (4 nm SW of Stanwood)	SSE		
21	Admiralty Bay (1 nm NE of Marrowstone Point)	NW		
22	Possession Sound (3 nm NE of Elliot Point)	SSW	NNW	
23	Useless Bay (3 nm NE of Point No Point)	S		
24	Hood Canal	SW		
25	Brown Point (mid-channel Hood Canal)	NE	SW	
26	Quatsap Point (mid-channel Hood Canal)	SW		
27	Annas Bay (south end Hood Canal)	NE		
28	Puget Sound (1.5 nm NW of Edwards Point)	NW		
29	Puget Sound (1 nm W of Edwards Point)	N	S	
30	Puget Sound (0.7 nm W of West Point)	N	S	
31	Puget Sound (north side of Maury I.)	N		
32	Henderson Bay (2 nm SSW of Elgin)	S		
33	Commencement Bay (1.5 nm SE of Neil Point)	NE		
34	Pt. Defiance	NNE		
35	The Narrows Bridge	SW		
36	Nisqually Reach (mid-channel E of Lyle Point)	NE		
37	Nisqually Reach (mid-channel W of Lyle Point)	NW		
38	Carr Inlet (2 nm E of South Head)	SE	NW	

The numbers on the chart (Figure C-2) correspond to the numbers in the Wave Height Tables. Wind directions are for TRUE directions from which wind blows, not magnetic directions. Note that no Wave Height Table is furnished for winds coming off the Pacific Ocean from the west at Cape Flattery. The reason for this is that an ocean wind fetch is quite variable, depending on the shape and size of the weather systems causing the winds. Just keep in mind that seas will be much higher on the open ocean for the same wind compared to wave heights in more protected waters. Values in Table C-1 were calculated using techniques from Shore Protection Manual - Vol. 1. U.S. Army Coastal Engineering Research Center, 1973.

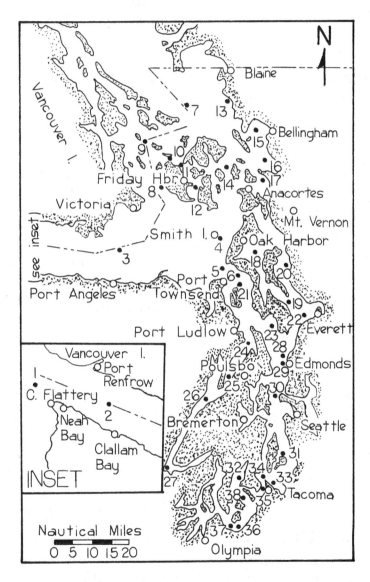

Figure C-2 Location points (numbered) for the wave height predictions of Table C-1.

TABLE C-1   WAVE HEIGHT PREDICTIONS FOR NORTHWEST WASHINGTON

Location 1 with ESE Wind

WIND DURATION (hr)

WIND (kn)	1	2	3	4	5	6	7	8
10	--	0.3	0.8	1.2	1.3	1.4	1.5	1.6	1.7
20	--	1.8	2.7	3.2	3.7	4.0	4.2		
30	--	3.2	4.6	5.7	6.6	7.0			
40	--	4.8	7.0	8.7	9.8				
50	--	6.7	9.8	12	13				
60	--	8.6	13	16					
70	--	11	16	19					

Location 2 with ESE Wind

WIND DURATION (hr)

WIND (kn)	1	2	3	4	5	6	7
10 --	0.3	0.8	1.2	1.3	1.4	1.5	1.7
20 --	1.8	2.7	3.2	3.7	4.0	4.1	
30 --	3.2	4.6	5.7	6.6	6.7		
40 --	4.8	7.0	8.7	9.6			
50 --	6.7	9.8	12				
60 --	8.6	13	16				
70 --	11	16	19				

Location 2 with WNW Wind

	1	2	3	4	5	6	7
10 --	0.3	0.8	1.2	1.3	1.4		
20 --	1.8	2.7	3.2				
30 --	3.2	4.6	5.2				
40 --	4.8	7.0	7.2				
50 --	6.7	9.4					
60 --	8.6	12					
70 --	11	14					

Location 3 with ESE Wind

	1	2	3	4	5	6	7
10 --	0.3	0.8	1.2	1.3	1.4	1.5	
20 --	1.8	2.7	3.2	3.6	3.7		
30 --	3.2	4.6	5.7	6.3			
40 --	4.8	7.0	8.6				
50 --	6.7	9.8	11				
60 --	8.6	13	14				
70 --	11	16	17				

Location 3 with WNW Wind

	1	2	3	4	5	6	7
10 --	0.3	0.8	1.2	1.3	1.4	1.5	1.6
20 --	1.8	2.7	3.2	3.6	3.9		
30 --	3.2	4.6	5.7	6.5			
40 --	4.8	7.0	8.7	9.0			
50 --	6.7	9.8	12				
60 --	8.6	13	15				
70 --	11	16	17				

Location 3 with NE Wind

	1	2	3	4	5	6	7
10 --	0.3	0.8	1.2	1.3	1.4	1.5	
20 --	1.8	2.7	3.2	3.7	4.0		
30 --	3.2	4.6	5.7	6.5			
40 --	4.8	7.0	8.7				
50 --	6.7	9.8	12				
60 --	8.6	13	15				
70 --	11	16	18				

Location 4 with N Wind

	1	2	3	4	5	6	7
10 --	0.3	0.8	1.2				
20 --	1.8	2.7	2.8				
30 --	3.2	4.3					
40 --	4.8	6.2					
50 --	6.7	8.0					
60 --	8.6	9.8					
70 --	11	12					

Location 4 with W Wind

	1	2	3	4	5	6	7
10 --	0.3	0.8	1.2	1.3	1.4	1.5	1.7
20 --	1.8	2.7	3.2	3.7	4.0		
30 --	3.2	4.6	5.7	6.5			
40 --	4.8	7.0	8.7	9.1			
50 --	6.7	9.8	12				
60 --	8.6	13	15				
70 --	11	16	18				

TABLE C-1 Continued.   WAVE HEIGHT PREDICTIONS FOR NORTHWEST WASHINGTON

Location 4 with NW Wind

WIND DURATION (hr)

WIND (kn)	1	2	3	4	5
10 —	0.3	0.8	1.2	1.3	1.4
20 —	1.8	2.7	3.2		
30 —	3.2	4.6	5.2		
40 —	4.8	7.0	7.2		
50 —	6.7	9.4			
60 —	8.6	12			
70 —	11	14			

Location 5 with N Wind

WIND DURATION (hr)

WIND (kn)	1	2	3	4	5	6	7
10 —	0.3	0.8	1.2				
20 —	1.8	2.7	2.8				
30 —	3.2	4.3					
40 —	4.8	6.2					
50 —	6.7	8.0					
60 —	8.6	10					
70 —	11	12					

Location 6 with SE Wind

	1	2	3	4	5
10 —	0.3	0.8	1.2		
20 —	1.8	2.7	2.8		
30 —	3.2	4.3			
40 —	4.8	6.2			
50 —	6.7	8.0			
60 —	8.6	10			
70 —	11	12			

Location 6 with WNW Wind

	1	2	3	4	5	6	7
10 —	0.3	0.8	1.2	1.3	1.4	1.5	1.6
20 —	1.8	2.7	3.2	3.6	4.0		
30 —	3.2	4.6	5.7	6.6			
40 —	6.7	7.0	8.7	9.3			
50 —	8.6	9.8	12				
60 —	8.6	13	15				
70 —	11	16	18				

Location 7 with SE Wind

	1	2	3	4	5
10 —	0.3	0.8	1.2	1.3	
20 —	1.8	2.7	3.0		
30 —	3.2	4.3	4.9		
40 —	4.8	6.8			
50 —	6.7	9.0			
60 —	8.6	11			
70 —	11	13			

Location 7 with WNW Wind

	1	2	3	4	5	6	7	8
10 —	0.3	0.8	1.2	1.3	1.4	1.5	1.6	1.7
20 —	1.8	2.7	3.2	3.7	4.0	4.2		
30 —	3.2	4.6	5.7	6.6	7.0			
40 —	4.8	7.0	8.7	9.8				
50 —	6.7	9.8	12	13				
60 —	8.6	13	16					
70 —	11	16	19					

Location 7 with NE Wind

	1	2	3	4	5
10 —	0.3	0.8	1.2	1.3	
20 —	1.8	2.7	3.0		
30 —	3.2	4.6	4.8		
40 —	4.8	6.8			
50 —	6.7	9.0			
60 —	8.6	11			
70 —	11	13			

Location 8 with SE Wind

	1	2	3	4	5	6	7
10 —	0.3	0.8	1.2	1.3	1.4	1.5	1.6
20 —	1.8	2.7	3.2	3.6	4.0		
30 —	3.2	4.6	5.7	6.6			
40 —	4.8	7.0	8.7	9.2			
50 —	6.7	9.8	12				
60 —	8.6	13	15				
70 —	11	16	18				

TABLE C-1 Continued.   WAVE HEIGHT PREDICTIONS FOR NORTHWEST WASHINGTON

Location 9 with SSE Wind

WIND DURATION (hr)

WIND (kn)	1	2	3	4	5	6
10	--	0.3	0.8	1.2	1.3	1.4	1.5
20	--	1.8	2.7	3.2	3.6	3.7	
30	--	3.2	4.6	5.7	6.3		
40	--	4.8	7.0	8.6			
50	--	6.7	9.8	11			
60	--	8.6	13	14			
70	--	11	16	17			

Location 9 with NE Wind

WIND DURATION (hr)

WIND (kn)	1	2	3	4	5	6
10	--	0.3	0.8	1.2			
20	--	1.8	2.7				
30	--	3.2	4.4				
40	--	4.8	6.0				
50	--	6.7	7.9				
60	--	8.6	9.7				
70	--	11	12				

Location 10 with NE Wind

	1	2	3	4	5	6
10 --	0.3	0.8	1.2	1.3	1.4	1.5
20 --	1.8	2.7	3.2	3.6		
30 --	3.2	4.6	5.7	6.0		
40 --	4.8	7.0	8.5			
50 --	6.7	9.8	11			
60 --	8.6	13	14			
70 --	11	16				

Location 10 with SE Wind

	1	2
10 --	0.3	0.6
20 --	1.8	1.9
30 --	3.0	
40 --	4.2	
50 --	5.3	
60 --	6.8	
70 --	7.9	

Location 10 with WNW Wind

	1	2	3
10 --	0.3	0.8	0.9
20 --	1.8	2.1	
30 --	3.2	3.5	
40 --	4.8		
50 --	6.3		
60 --	7.7		
70 --	9.3		

Location 11 with SE Wind

	1	2
10 --	0.3	0.6
20 --	1.7	
30 --	2.8	
40 --	3.8	
50 --	5.0	
60 --	6.1	
70 --	7.4	

Location 11 with NW Wind

	1	2	3
10 --	0.3	0.8	1.1
20 --	1.8	2.4	
30 --	3.2	3.8	
40 --	4.8	5.5	
50 --	6.7	7.0	
60 --	8.6		
70 --	11		

Location 12 with SSE Wind

	1	2	3	4	5	6
10 --	0.3	0.8	1.2	1.3	1.4	1.5
20 --	1.8	2.7	3.2	3.6	3.7	
30 --	3.2	4.6	5.7	6.3		
40 --	4.8	7.0	8.6			
50 --	6.7	9.8	11			
60 --	8.6	13	14			
70 --	11	16	17			

Location 13 with S Wind

	1	2	3	4	5
10 --	0.3	0.8	1.2	1.3	1.4
20 --	1.8	2.7	3.2	3.3	
30 --	3.2	4.6	5.3		
40 --	4.8	7.0	7.6		
50 --	6.7	9.8	9.9		
60 --	8.6	12			
70 --	11	15			

Location 14 with N Wind

	1	2	3
10 --	0.3	0.8	1.2
20 --	1.8	2.5	
30 --	3.2	4.0	
40 --	4.8	5.6	
50 --	6.7	7.2	
60 --	8.6	9.0	
70 --	11		

Location 14 with S Wind

	1	2	3	4	5
10 --	0.3	0.8	1.2	1.3	1.4
20 --	1.8	2.7	3.2		
30 --	3.2	4.6	5.2		
40 --	4.8	7.0	7.2		
50 --	6.7	9.4			
60 --	8.6	12			
70 --	11	14			

TABLE C-1 Continued.   WAVE HEIGHT PREDICTIONS FOR NORTHWEST WASHINGTON

Location 15 with SSE Wind

WIND (kn)	WIND DURATION (hr)				
	1	2	3	4	5
10	--	0.3	0.8	1.2	
20	--	1.8	2.5		
30	--	3.2	4.0		
40	--	4.8	5.6		
50	--	6.7	7.2		
60	--	8.6	9.0		
70	--	11			

Location 16 with NWW Wind

WIND (kn)	WIND DURATION (hr)				
	1	2	3	4	5
10	--	0.3	0.8	1.1	
20	--	1.8	2.4		
30	--	3.2	3.8		
40	--	4.8	5.5		
50	--	6.7	7.0		
60	--	8.6			
70	--	11			

Location 17 with NW Wind

	1	2	3	4	5	6
10	--	0.3	0.8	1.2	1.3	1.4	1.5
20	--	1.8	2.7	3.0	3.5		
30	--	3.2	4.6	5.7			
40	--	4.8	7.0	8.0			
50	--	6.7	9.8	11			
60	--	8.6	13				
70	--	11	16				

Location 18 with SSE Wind

	1	2
10	--	0.3	0.6
20	--	1.7	
30	--	2.8	
40	--	3.8	
50	--	5.0	
60	--	6.1	
70	--	7.4	

Location 19 with NE Wind

	1	2	3
10	--	0.3	0.7	
20	--	1.8	2.0	
30	--	3.2		
40	--	4.4		
50	--	5.8		
60	--	7.0		
70	--	8.4		

Location 19 with SE Wind

	1	2	3
10	--	0.3	0.8	0.9
20	--	1.8	2.1	
30	--	3.2	3.5	
40	--	4.8		
50	--	6.3		
60	--	7.7		
70	--	9.3		

Location 20 with SSE Wind

	1	2	3
10	--	0.3	0.8	1.1
20	--	1.8	2.4	
30	--	3.2	3.8	
40	--	4.8	5.5	
50	--	6.7	7.0	
60	--	8.6		
70	--	11		

Location 21 with NW Wind

	1	2	3	4	5	6	7
10	--	0.3	0.8	1.2	1.3	1.4	1.5	1.7
20	--	1.8	2.7	3.2	3.7	4.0		
30	--	3.2	4.6	5.7	6.5			
40	--	4.8	7.0	8.7	9.1			
50	--	6.7	9.8	12				
60	--	8.6	13	15				
70	--	11	16	18				

Location 22 with SSW Wind

	1	2	3
10	--	0.3	0.8	1.2
20	--	1.8	2.7	2.8
30	--	3.2	4.3	
40	--	4.8	6.2	
50	--	6.7	8.0	
60	--	8.6	9.8	
70	--	11	12	

Location 22 with NNW Wind

	1	2	3
10	--	0.3	0.8	0.9
20	--	1.8	2.3	
30	--	3.2	3.6	
40	--	4.8	5.0	
50	--	6.6		
60	--	8.1		
70	--	9.7		

Location 23 with S Wind

	1	2	3
10	--	0.3	0.8	0.9
20	--	1.8	2.3	
30	--	3.2	3.6	
40	--	4.8	5.0	
50	--	6.6		
60	--	8.1		
70	--	9.7		

Location 24 with SW Wind

	1	2
10	--	0.3	0.6
20	--	1.8	1.9
30	--	3.0	
40	--	4.2	
50	--	5.3	
60	--	6.8	
70	--	7.9	

TABLE C-1 Continued. WAVE HEIGHT PREDICTIONS FOR NORTHWEST WASHINGTON

Location 25 with NE Wind

WIND DURATION (hr)

WIND (kn)	1	2	3	4	5
10	—	0.3	0.6		
20	—	1.8	1.9		
30	—	3.0			
40	—	4.2			
50	—	5.3			
60	—	6.8			
70	—	7.9			

Location 25 with SW Wind

WIND DURATION (hr)

WIND (kn)	1	2	3	4	5
10	—	0.3	0.6		
20	—	1.8	1.9		
30	—	3.0			
40	—	4.2			
50	—	5.3			
60	—	6.8			
70	—	7.9			

Location 26 with SW Wind

	1	2
10	— 0.3	0.6
20	— 1.8	1.9
30	— 3.0	
40	— 4.2	
50	— 5.3	
60	— 6.8	
70	— 7.9	

Location 27 with NE Wind

	1	2
10	— 0.3	0.6
20	— 1.8	1.9
30	— 3.0	
40	— 4.2	
50	— 5.3	
60	— 6.8	
70	— 7.9	

Location 28 with NW Wind

	1	2	3	4	5	6
10	— 0.3	0.8	1.2	1.3	1.4	1.5
20	— 1.8	2.7	3.2	3.7	3.9	
30	— 3.2	4.6	5.7	6.1		
40	— 4.8	7.0	8.7			
50	— 6.7	9.8	11			
60	— 8.6	13	14			
70	— 11	16	17			

Location 29 with N Wind

	1	2	3
10	— 0.3	0.8	1.1
20	— 1.8	2.4	
30	— 3.2	3.8	
40	— 4.8	5.5	
50	— 6.7	7.0	
60	— 8.6		
70	— 11		

Location 29 with S Wind

	1	2	3
10	— 0.3	0.8	0.9
20	— 1.8	2.1	
30	— 3.2	3.5	
40	— 4.8		
50	— 6.3		
60	— 7.7		
70	— 9.3		

Location 30 with N Wind

	1	2	3	4
10	— 0.3	0.8	1.2	
20	— 1.8	2.7		
30	— 3.2	4.2		
40	— 4.8	5.8		
50	— 6.7	7.6		
60	— 8.6	9.4		
70	— 11			

Location 30 with S Wind

	1	2	3
10	— 0.3	0.8	0.9
20	— 1.8	2.1	
30	— 3.2	3.5	
40	— 4.8		
50	— 6.3		
60	— 7.7		
70	— 9.3		

Location 31 with N Wind

	1	2	3
10	— 0.3	0.8	1.2
20	— 1.8	2.7	2.8
30	— 3.2	4.3	
40	— 4.8	6.2	
50	— 6.7	8.0	
60	— 8.6	9.8	
70	— 11	12	

Location 32 with S Wind

	1	2
10	— 0.3	0.6
20	— 1.8	1.9
30	— 3.0	
40	— 4.2	
50	— 5.3	
60	— 6.8	
70	— 7.9	

Location 33 with NE Wind

	1	2
10	— 0.3	0.7
20	— 1.8	2.0
30	— 3.2	
40	— 4.4	
50	— 5.8	
60	— 7.0	
70	— 8.4	

TABLE C-1 Continued. WAVE HEIGHT PREDICTIONS FOR NORTHWEST WASHINGTON

Location 34 with NNE Wind

WIND (kn)	WIND DURATION (hr)				
	1	2	3	4	5
10	0.3	0.8	1.3		
20	1.8	2.7			
30	3.2	4.1			
40	4.7	5.7			
50	6.7	7.5			
60	8.7	9.2			
70	11				

Location 35 with SW Wind

WIND (kn)	WIND DURATION (hr)				
	1	2	3	4	5
10	0.3	0.7			
20	1.8	2.0			
30	3.2				
40	4.4				
50	5.8				
60	7.0				
70	8.4				

Location 36 with NE Wind	Location 37 with NW Wind	Location 38 with SE Wind		Location 38 with NW Wind	
1	1	1	2	1	2
10 -- 0.4	10 -- 0.7	10 -- 0.3	0.6	10 -- 0.3	0.6
20 -- 1.4	20 -- 1.7	20 -- 1.8	1.9	20 -- 1.8	1.9
30 -- 2.2	30 -- 2.8	30 -- 3.0		30 -- 3.0	
40 -- 3.1	40 -- 3.9	40 -- 4.2		40 -- 4.2	
50 -- 4.0	50 -- 5.0	50 -- 5.3		50 -- 5.3	
60 -- 5.0	60 -- 6.3	60 -- 6.8		60 -- 6.8	
70 -- 5.9	70 -- 7.5	70 -- 7.9		70 -- 7.9	

APPENDIX D

WAVE HEIGHT TABLES FOR THE OPEN OCEAN

These tables apply to open ocean areas where the fetch is not limited in <u>width</u> by nearby terrain and where the water depth is over 150 feet. It is assumed that water currents are less than one knot.

It is beyond the scope of this book to go into the details of wave forecasting for the open ocean because it is necessary to have weather maps to determine wind fetch, speed, and duration. These maps can only be received aboard a vessel at sea by use of radiofacsimile. The tables are included here to indicate the much greater magnitude of the waves on the open ocean as compared to protected waters. Figure D-1 illustrates three factors that can limit fetch and hence the height of ocean waves.

The tables can be used to determine significant WIND WAVE heights in FEET once the average WIND SPEED in KNOTS, WIND DURATION in HOURS, and the WIND FETCH in NAUTICAL MILES have been determined. Significant Wave Height (discussed in Chapter 2) is a statistical term representing the average height of the highest one third of the waves. It is a value higher than the value obtained by averaging all the wave heights, but it is smaller than the highest waves observed. Experienced mariners, when reporting wave heights, typically report values very close to the Significant Wave Height. Once the Significant Wave Height is known, estimates of the frequency and heights of other waves to be expected can be made by using Table 2-1 in Chapter 2.

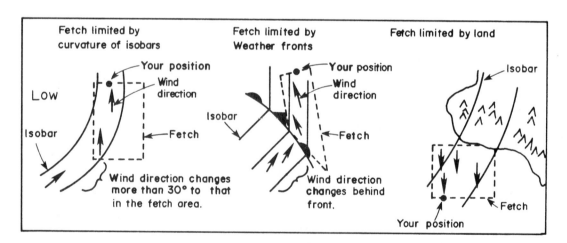

Figure D-1 Examples of fetch limitations. Note that fetch should not be confused with sea room. It is not always the physical extent of the body of water that limits fetch and hence wave height. A squall in the open ocean may have strong winds but a fetch of only a few miles. Squall waves rarely reflect the potential of the wind strength. On the other hand, the only moderate strengths of the tradewinds raise memorable waves because their fetch and duration are very long.

TABLE D-1   WAVE HEIGHT PREDICTIONS FOR THE OPEN OCEAN

WIND SPEED 20 knots

WIND DURATION (hr)	FETCH (nm)														
	2	5	10	20	30	40	50	100	150	200	300	400	600	800	1000
1	1.5	1.7	1.7	1.7	1.7	1.7	1.7	1.7	1.7	1.7	1.7	1.7	1.7	1.7	1.7
2		2.2	2.6	2.6	2.6	2.6	2.6	2.6	2.6	2.6	2.6	2.6	2.6	2.6	2.6
3			2.8	3.2	3.2	3.2	3.2	3.2	3.2	3.2	3.2	3.2	3.2	3.2	3.2
6				3.7	4.4	4.4	4.4	4.4	4.4	4.4	4.4	4.4	4.4	4.4	4.4
12						4.9	5.1	6.0	6.0	6.0	6.0	6.0	6.0	6.0	6.0
24								6.4	7.1	7.5	7.5	7.5	7.5	7.5	7.5
48	1.5	2.2	2.8	3.7	4.4	4.9	5.1	6.4	7.1	7.8	8.5	9.0	10	10	10

WIND SPEED 30 knots

WIND DURATION (hr)	FETCH (nm)														
	2	5	10	20	30	40	50	100	150	200	300	400	600	800	1000
1	2.4	3.2	3.2	3.2	3.2	3.2	3.2	3.2	3.2	3.2	3.2	3.2	3.2	3.2	3.2
2		3.5	4.6	4.6	4.6	4.6	4.6	4.6	4.6	4.6	4.6	4.6	4.6	4.6	4.6
3				5.9	5.9	5.9	5.9	5.9	5.9	5.9	5.9	5.9	5.9	5.9	5.9
6				6.0	7.0	8.0	8.1	8.1	8.1	8.1	8.1	8.1	8.1	8.1	8.1
12							8.8	11	11	11	11	11	11	11	11
24									13	14	15	15	15	15	15
48	2.4	3.5	4.6	6.0	7.0	8.0	8.8	11	13	14	16	17	18	18	18

WIND SPEED 40 knots

WIND DURATION (hr)	FETCH (nm)														
	2	5	10	20	30	40	50	100	150	200	300	400	600	800	1000
1	3.3	4.9	4.9	4.9	4.9	4.9	4.9	4.9	4.9	4.9	4.9	4.9	4.9	4.9	4.9
2			6.5	7.0	7.0	7.0	7.0	7.0	7.0	7.0	7.0	7.0	7.0	7.0	7.0
3				8.6	8.9	8.9	8.9	8.9	8.9	8.9	8.9	8.9	8.9	8.9	8.9
6					10	11	12	13	13	13	13	13	13	13	13
12								16	17	17	17	17	17	17	17
24									18	20	24	24	24	24	24
48	3.3	4.9	6.5	8.6	10	11	12	16	18	20	24	26	29	30	30

TABLE D-1 Continued.   WAVE HEIGHT PREDICTIONS FOR THE OPEN OCEAN

WIND SPEED 50 knots

WIND DURATION (hr)	FETCH (nm)														
	2	5	10	20	30	40	50	100	150	200	300	400	600	800	1000
1	4.3	6.2	6.7	6.7	6.7	6.7	6.7	6.7	6.7	6.7	6.7	6.7	6.7	6.7	6.7
2		8.2	9.8	9.8	9.8	9.8	9.8	9.8	9.8	9.8	9.8	9.8	9.8	9.8	9.8
3			11	12	12	12	12	12	12	12	12	12	12	12	12
6				13	15	16	18	18	18	18	18	18	18	18	18
12								22	25	25	25	25	25	25	25
24										27	32	34	34	34	34
48	4.3	6.2	8.2	11	13	15	16	22	25	27	32	35	40	43	45

WIND SPEED 60 knots

WIND DURATION (hr)	FETCH (nm)														
	2	5	10	20	30	40	50	100	150	200	300	400	600	800	1000
1	5.3	7.8	8.8	8.8	8.8	8.8	8.8	8.8	8.8	8.8	8.8	8.8	8.8	8.8	8.8
2			10	13	13	13	13	13	13	13	13	13	13	13	13
3			14	16	16	16	16	16	16	16	16	16	16	16	16
6					18	20	23	23	23	23	23	23	23	23	23
12								27	31	33	33	33	33	33	33
24										35	40	45	45	45	45
48	5.3	7.8	10	14	16	18	20	27	31	35	40	45	51	55	60

WIND SPEED 70 knots

WIND DURATION (hr)	FETCH (nm)														
	2	5	10	20	30	40	50	100	150	200	300	400	600	800	1000
1	6.3	9.2	11	11	11	11	11	11	11	11	11	11	11	11	11
2			12	16	16	16	16	16	16	16	16	16	16	16	16
3				17	20	20	20	20	20	20	20	20	20	20	20
6						23	24	28	28	28	28	28	28	28	28
12								32	37	40	40	40	40	40	40
24										42	48	54	57	57	57
48	6.3	9.2	12	17	20	23	24	32	37	42	48	54	62	69	72

APPENDIX E

WATER TEMPERATURES IN NORTHWEST WASHINGTON

TABLE E-1 WATER TEMPERATURES IN WESTERN WASHINGTON[a]

Location[b]	Average Surface Water Temperature in °F											
	Jan	Feb	Mar	Apr	May	Jun	Jul	Aug	Sep	Oct	Nov	Dec
Seattle[c] (39)	47	47	47	48	51	53	56	56	55	54	52	49
Bremerton (3)	46	45	46	48	53	57	59	60	59	55	51	48
Tacoma (2)	45	45	45	47	52	56	58	58	55	53	50	48
Olympia (4)	44	44	45	50	54	58	61	62	59	55	51	47
Everett (2)	45	44	45	49	55	57	62	61	58	53	50	47
Port Townsend (3)	44	44	44	46	48	50	52	52	51	50	48	46
Anacortes (5)	44	44	45	47	49	51	53	53	53	51	48	46
Friday Harbor (18)	45	45	45	47	49	50	52	52	51	50	48	47
Blaine (2)	41	44	44	48	54	59	61	61	57	51	48	44
Port Angeles (2)	45	46	47	49	51	52	53	54	52	50	50	47
Neah Bay (25)	45	45	46	49	51	53	53	53	52	51	49	47
Aberdeen (2)	43	45	45	51	58	63	64	65	62	56	50	45
Raymond (8)	44	45	48	54	60	64	69	68	65	58	49	46
Tokeland (5)	45	44	48	51	56	61	63	64	60	56	50	47

[a] Values are from *Surface Water Temperature and Salinity -- Pacific Coast Pub. 31-1* published by the National Ocean Service.

[b] Numbers following location are the years of record.

[c] At Elliott Bay.

Water temperatures may vary widely from the averages shown in the table. For example, the following table shows the lowest and highest observed water temperatures at those stations where observations have been made for a number of years.

LOCATION	MINIMUM (°F)	MAXIMUM (°F)
Seattle (Elliott Bay)	37	70
Neah Bay	36	64
Friday Harbor	41	61

APPENDIX F

HYPOTHERMIA

Hypothermia (lowered deep-body temperature) is a definite threat to survival in the cool waters of northwestern Washington and should be considered when making weather decisions that could affect the safe operation of a boat. Very extensive research at the University of Victoria, Canada, resulted in many findings about hypothermia which are summarized below.

1. Unconsciousness can occur when the internal body temperature of the victim falls to about 90°F (32°C). Heart failure is the usual cause of death when the core body temperature reaches about 85°F (30°C) or below.

2. Skin tissues cool very rapidly, but it takes 10-15 minutes before the temperature of the heart and brain begin to cool.

3. Predicted survival times of an average adult human wearing a standard life jacket and light clothing are shown in Figure 7-1. Extra body fat increases survival time. Small body size, especially for children, decreases survival time.

4. Swimming causes a person to cool 35% faster than holding still due to increased blood circulation to the arms and legs. However, if the shore is within one mile and the water temperature is near 50°F (10°C), the average person in a life jacket can probably swim the distance before being overcome by hypothermia.

5. If you fall in the water without a life preserver, your survival time decreases. Table F-1 shows predicted survival times in 50°F (10°C) water under various situations in which an average adult is wearing a cotton shirt, pants, socks, and tennis shoes.

6. Death can occur in cold water even without hypothermia. Immersion in cold water (especially if sudden) causes a rapid rise in heart beat and blood pressure which occasionally results in heart attacks or ruptured blood vessels. Cold shock causes immediate hyperventilation (over-breathing) and has been known to cause even expert swimmers to aspirate water and drown, particularly when people have plunged underwater or are in large waves. Hyperventilation may cause changes in blood chemistry leading to unconsciousness and possible drowning.

7. You do not have to fall in the water to suffer hypothermia. Cold winds, being wet from spray or rain, or a combination of the two can also cause hypothermia. The author recalls an incident in which two young people died of hypothermia near Bremerton. They departed one afternoon in a skiff without proper waterproof clothing and suffered fatal hypothermia that night due to strong winds, temperatures around 40°F, rain, and spray because they were unable to make it back to shore.

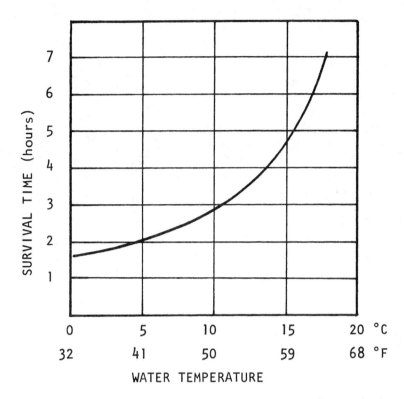

Figure F-1 Adult survival times in water wearing life jacket and light clothing.

TABLE F-1 PREDICTED SURVIVAL TIMES UNDER VARIOUS CONDITIONS[a]

Situation	Survival Time (hr)
No Flotation:	
Drownproofing	1.5
Treading water	2.0
With Flotation:	
Swimming	2.0
Holding Still	2.7
Heat Escape Lessening Position (HELP)[b]	4.0
Huddle (Two or more people)	4.0
UVIC Thermofloat Jacket[c]	9.5

[a] Water temperature 50°F; person dressed in light clothes.

[b] H E L P is a procedure whereby the sides of the arms are held tightly against the side of the chest, and thighs are raised to close off the groin area.

[c] This is a floater jacket with a specially designed insulating foam flap that can be pulled down from inside the coat to seal off the upper leg area.

APPENDIX G

WIND CHILL

How cold you feel depends not only on the dry bulb air temperature you measure on your thermometer, but how active you are, and what type of clothing you have on. A wind chill chart, such as the one here, takes into account the effects of wind in cooling a person off. If, for example, the air temperature is 45°F and the wind is 15 mph, you will feel like it is actually 29°F out and will likely be dressed accordingly. The wind chill chart only applies if you are not wet. Depending on the relative humidity, you will feel considerably colder if you are wet and the wind is blowing because the evaporation of the water will carry body heat rapidly away.

In a moving boat, the direction you are traveling in relation to the wind that is blowing will create an apparent wind -- a wind that may be greater or smaller than the natural wind that is blowing. It is the apparent wind you should use in the above table. If the boat is not moving, then the apparent wind is the same as the true wind.

Example:

13-kn true wind	13-kn boat speed, headed into the wind	Apparent wind is 26 knots Chill temperature at 50°F is 29°F.
13-kn true wind	13-kn boat speed, headed down wind	Apparent wind is 0 knots Chill temperature at 50°F is 50°F.

Thus, it would feel 21°F colder when heading into the wind in this case than when you are heading in the same direction as the natural wind.

TABLE G-1   WIND CHILL TEMPERATURE VERSUS WIND SPEED

Wind Speed		Air Temperature °F																			
mph	kn	70	65	60	55	50	45	40	35	30	25	20	15	10	5	0	-5	-10	-15	-20	-25	-30
5	4	64	62	59	51	46	43	37	32	27	22	16	11	6	0	-5	-10	-15	-21	-26	-31	-36
10	9	62	60	53	45	40	34	28	22	16	10	3	-3	-9	-15	-22	-27	-34	-40	-46	-52	-58
15	13	60	56	48	41	35	29	23	16	9	2	-5	-11	-18	-25	-31	-38	-45	-51	-58	-65	-72
20	17	59	54	45	39	32	26	19	12	4	-3	-10	-17	-24	-31	-39	-46	-53	-60	-67	-74	-81
25	22	59	51	44	36	30	23	16	8	1	-7	-15	-22	-29	-36	-44	-51	-59	-66	-74	-81	-88
30	26	59	50	44	35	29	21	13	6	-2	-10	-18	-25	-33	-41	-49	-56	-64	-71	-79	-86	-93
35	30	59	50	44	34	27	20	12	4	-4	-12	-20	-27	-35	-43	-52	-58	-67	-74	-82	-89	-97
40	35	58	50	44	33	26	19	11	3	-5	-13	-21	-29	-37	-45	-53	-60	-69	-76	-84	-92	-100

Considerable danger from freezing of exposed flesh.

Very great danger from freezing of exposed flesh.

APPENDIX H

CONVERSION TABLES FOR PRESSURES AND TEMPERATURES

TABLE H-1
Conversion Table for Millibars, Inches of Mercury, and Millimeters of Mercury

Millibars	Inches	Millimeters	Millibars	Inches	Millimeters	Millibars	Inches	Millimeters
900	26.58	675.1	960	28.35	720.1	1020	30.12	765.1
901	26.61	675.8	961	28.38	720.8	1021	30.15	765.8
902	26.64	676.6	962	28.41	721.6	1022	30.18	766.6
903	26.67	677.3	963	28.44	722.3	1023	30.21	767.3
904	26.70	678.1	964	28.47	723.1	1024	30.24	768.1
905	26.72	678.8	965	28.50	723.8	1025	30.27	768.8
906	26.75	679.6	966	28.53	724.6	1026	30.30	769.6
907	26.78	680.3	967	28.56	725.3	1027	30.33	770.3
908	26.81	681.1	968	28.58	726.1	1028	30.36	771.1
909	26.84	681.8	969	28.61	726.8	1029	30.39	771.8
910	26.87	682.6	970	28.64	727.6	1030	30.42	772.6
911	26.90	683.3	971	28.67	728.3	1031	30.45	773.3
912	26.93	684.1	972	28.70	729.1	1032	30.47	774.1
913	26.96	684.8	973	28.73	729.8	1033	30.50	774.8
914	26.99	685.6	974	28.76	730.6	1034	30.53	775.6
915	27.02	686.3	975	28.79	731.3	1035	30.56	776.3
916	27.05	687.1	976	28.82	732.1	1036	30.59	777.1
917	27.08	687.8	977	28.85	732.8	1037	30.62	777.8
918	27.11	688.6	978	28.88	733.6	1038	30.65	778.6
919	27.14	689.3	979	28.91	734.3	1039	30.68	779.3
920	27.17	690.1	980	28.94	735.1	1040	30.71	780.1
921	27.20	690.8	981	28.97	735.8	1041	30.74	780.8
922	27.23	691.6	982	29.00	736.6	1042	30.77	781.6
923	27.26	692.3	983	29.03	737.3	1043	30.80	782.3
924	27.29	693.1	984	29.06	738.1	1044	30.83	783.1
925	27.32	693.8	985	29.09	738.8	1045	30.86	783.8
926	27.34	694.6	986	29.12	739.6	1046	30.89	784.6
927	27.37	695.3	987	29.15	740.3	1047	30.92	785.3
928	27.40	696.1	988	29.18	741.1	1048	30.95	786.1
929	27.43	696.8	989	29.21	741.8	1049	30.98	786.8
930	27.46	697.6	990	29.23	742.6	1050	31.01	787.6
931	27.49	698.3	991	29.26	743.3	1051	31.04	788.3
932	27.52	699.1	992	29.29	744.1	1052	31.07	789.1
933	27.55	699.8	993	29.32	744.8	1053	31.10	789.8
934	27.58	700.6	994	29.35	745.6	1054	31.12	790.6
935	27.61	701.3	995	29.38	746.3	1055	31.15	791.3
936	27.64	702.1	996	29.41	747.1	1056	31.18	792.1
937	27.67	702.8	997	29.44	747.8	1057	31.21	792.8
938	27.70	703.6	998	29.47	748.6	1058	31.24	793.6
939	27.73	704.3	999	29.50	749.3	1059	31.27	794.3
940	27.76	705.1	1000	29.53	750.1	1060	31.30	795.1
941	27.79	705.8	1001	29.56	750.8	1061	31.33	795.8
942	27.82	706.6	1002	29.59	751.6	1062	31.36	796.6
943	27.85	707.3	1003	29.62	752.3	1063	31.39	797.3
944	27.88	708.1	1004	29.65	753.1	1064	31.42	798.1
945	27.91	708.8	1005	29.68	753.8	1065	31.45	798.8
946	27.94	709.6	1006	29.71	754.6	1066	31.48	799.6
947	27.96	710.3	1007	29.74	755.3	1067	31.51	800.3
948	27.99	711.1	1008	29.77	756.1	1068	31.54	801.1
949	28.02	711.8	1009	29.80	756.8	1069	31.57	801.8
950	28.05	712.6	1010	29.83	757.6	1070	31.60	802.6
951	28.08	713.3	1011	29.85	758.3	1071	31.63	803.3
952	28.11	714.1	1012	29.88	759.1	1072	31.66	804.1
953	28.14	714.8	1013	29.91	759.8	1073	31.69	804.8
954	28.17	715.6	1014	29.94	760.6	1074	31.72	805.6
955	28.20	716.3	1015	29.97	761.3	1075	31.74	806.3
956	28.23	717.1	1016	30.00	762.1	1076	31.77	807.1
957	28.26	717.8	1017	30.03	762.8	1077	31.80	807.8
958	28.29	718.6	1018	30.06	763.6	1078	31.83	808.6
959	28.32	719.3	1019	30.09	764.3	1079	31.86	809.3
960	28.35	720.1	1020	30.12	765.1	1080	31.89	810.1

TABLE H-2

Conversion Tables for Thermometer Scales

F = Fahrenheit, C = Celsius (centigrade), K = Kelvin

F°	C°	K°	F°	C°	K°	C°	F°	K°	K°	F°	C°
−20	−28.9	244.3	+40	+4.4	277.6	−25	−13.0	248.2	250	−9.7	−23.2
19	28.3	244.8	41	5.0	278.2	24	11.2	249.2	251	7.9	22.2
18	27.8	245.4	42	5.6	278.7	23	9.4	250.2	252	6.1	21.2
17	27.2	245.9	43	6.1	279.3	22	7.6	251.2	253	4.3	20.2
16	26.7	246.5	44	6.7	279.8	21	5.8	252.2	254	2.5	19.2
−15	−26.1	247.0	+45	+7.2	280.4	−20	−4.0	253.2	255	−0.7	−18.2
14	25.6	247.6	46	7.8	280.9	19	2.2	254.2	256	+1.1	17.2
13	25.0	248.2	47	8.3	281.5	18	−0.4	255.2	257	2.9	16.2
12	24.4	248.7	48	8.9	282.0	17	+1.4	256.2	258	4.7	15.2
11	23.9	249.3	49	9.4	282.6	16	3.2	257.2	259	6.5	14.2
−10	−23.3	249.8	+50	+10.0	283.2	−15	+5.0	258.2	260	+8.3	−13.2
9	22.8	250.4	51	10.6	283.7	14	6.8	259.2	261	10.1	12.2
8	22.2	250.9	52	11.1	284.3	13	8.6	260.2	262	11.9	11.2
7	21.7	251.5	53	11.7	284.8	12	10.4	261.2	263	13.7	10.2
6	21.1	252.0	54	12.2	285.4	11	12.2	262.2	264	15.5	9.2
−5	−20.6	252.6	+55	+12.8	285.9	−10	+14.0	263.2	265	+17.3	−8.2
4	20.0	253.2	56	13.3	286.5	9	15.8	264.2	266	19.1	7.2
3	19.4	253.7	57	13.9	287.0	8	17.6	265.2	267	20.9	6.2
2	18.9	254.3	58	14.4	287.6	7	19.4	266.2	268	22.7	5.2
−1	18.3	254.8	59	15.0	288.2	6	21.2	267.2	269	24.5	4.2
0	−17.8	255.4	+60	+15.6	288.7	−5	+23.0	268.2	270	+26.3	−3.2
+1	17.2	255.9	61	16.1	289.3	4	24.8	269.2	271	28.1	2.2
2	16.7	256.5	62	16.7	289.8	3	26.6	270.2	272	29.9	1.2
3	16.1	257.0	63	17.2	290.4	2	28.4	271.2	273	31.7	−0.2
4	15.6	257.6	64	17.8	290.9	−1	30.2	272.2	274	33.5	+0.8
+5	−15.0	258.2	+65	+18.3	291.5	0	+32.0	273.2	275	+35.3	+1.8
6	14.4	258.7	66	18.9	292.0	+1	33.8	274.2	276	37.1	2.8
7	13.9	259.3	67	19.4	292.6	2	35.6	275.2	277	38.9	3.8
8	13.3	259.8	68	20.0	293.2	3	37.4	276.2	278	40.7	4.8
9	12.8	260.4	69	20.6	293.7	4	39.2	277.2	279	42.5	5.8
+10	−12.2	260.9	+70	+21.1	294.3	+5	+41.0	278.2	280	+44.3	+6.8
11	11.7	261.5	71	21.7	294.8	6	42.8	279.2	281	46.1	7.8
12	11.1	262.0	72	22.2	295.4	7	44.6	280.2	282	47.9	8.8
13	10.6	262.6	73	22.8	295.9	8	46.4	281.2	283	49.7	9.8
14	10.0	263.2	74	23.3	296.5	9	48.2	282.2	284	51.5	10.8
+15	−9.4	263.7	+75	+23.9	297.0	+10	+50.0	283.2	285	+53.3	+11.8
16	8.9	264.3	76	24.4	297.6	11	51.8	284.2	286	55.1	12.8
17	8.3	264.8	77	25.0	298.2	12	53.6	285.2	287	56.9	13.8
18	7.8	265.4	78	25.6	298.7	13	55.4	286.2	288	58.7	14.8
19	7.2	265.9	79	26.1	299.3	14	57.2	287.2	289	60.5	15.8
+20	−6.7	266.5	+80	+26.7	299.8	+15	+59.0	288.2	290	+62.3	+16.8
21	6.1	267.0	81	27.2	300.4	16	60.8	289.2	291	64.1	17.8
22	5.6	267.6	82	27.8	300.9	17	62.6	290.2	292	65.9	18.8
23	5.0	268.2	83	28.3	301.5	18	64.4	291.2	293	67.7	19.8
24	4.4	268.7	84	28.9	302.0	19	66.2	292.2	294	69.5	20.8
+25	−3.9	269.3	+85	+29.4	302.6	+20	+68.0	293.2	295	+71.3	+21.8
26	3.3	269.8	86	30.0	303.2	21	69.8	294.2	296	73.1	22.8
27	2.8	270.4	87	30.6	303.7	22	71.6	295.2	297	74.9	23.8
28	2.2	270.9	88	31.1	304.3	23	73.4	296.2	298	76.7	24.8
29	1.7	271.5	89	31.7	304.8	24	75.2	297.2	299	78.5	25.8
+30	−1.1	272.0	+90	+32.2	305.4	+25	+77.0	298.2	300	+80.3	+26.8
31	0.6	272.6	91	32.8	305.9	26	78.8	299.2	301	82.1	27.8
32	0.0	273.2	92	33.3	306.5	27	80.6	300.2	302	83.9	28.8
33	+0.6	273.7	93	33.9	307.0	28	82.4	301.2	303	85.7	29.8
34	1.1	274.3	94	34.4	307.6	29	84.2	302.2	304	87.5	30.8
+35	+1.7	274.8	+95	+35.0	308.2	+30	+86.0	303.2	305	+89.3	+31.8
36	2.2	275.4	96	35.6	308.7	31	87.8	304.2	306	91.1	32.8
37	2.8	275.9	97	36.1	309.3	32	89.6	305.2	307	92.9	33.8
38	3.3	276.5	98	36.7	309.8	33	91.4	306.2	308	94.7	34.8
39	3.9	277.0	99	37.2	310.4	34	93.2	307.2	309	96.5	35.8
+40	+4.4	277.6	+100	+37.8	310.9	+35	+95.0	308.2	310	+98.3	+36.8

APPENDIX I

CLOUD IDENTIFICATION

Only the basic cloud types are shown here, but these are sufficient for most purposes and should be of help in reading the sky for weather signs. Cloud identification is a long subject in itself; entire books are devoted to the topic. Some of the books listed in Appendix J have excellent pictures and detailed discussions of clouds. The identification of a cloud is based on its appearance and the height of its base above the earth's surface. In latitudes outside the tropics and polar regions, these cloud base heights are used:

Low Clouds	Ground to 6,500 ft.
Middle-height Clouds	6,500 to 23,000 ft.
High Clouds	16,500 to 45,000 ft.

Elevations of middle and high clouds are lower in the polar regions and somewhat higher in the tropics, but the elevation range of low clouds is the same everywhere. Note that the middle and high cloud ranges overlap, meaing that both high clouds and middle-height clouds can be found in the overlap range.

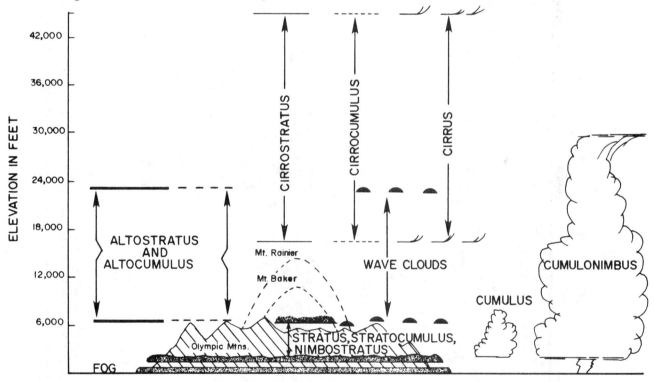

Figure I-1 Cloud types compared to terrain. By comparing the cloud base with the elevation of the terrain, it is usually possible to determine whether or not the cloud belongs to the low, middle, or high category. For example, altocumulus or altostratus clouds would be above most peaks of the Olympic and Cascade mountains, but not necessarily above Mt. Rainier or other peaks higher than 6,500 feet.

WAVE CLOUDS (lenticular altocumulus)—A middle cloud; almond or saucer shaped with distinct edges. Caused by winds over about 50 kn. blowing across mountains. Series of wave clouds may be observed downwind from a mountain and/or as a "cap" on the mountain peak. Stratocumulus and cirrocumulus may also be formed into wave clouds.

CIRRUS—A high cloud; fibrous and wispy with blue sky visible between cloud filaments. Sometimes called "mare's tails."

CIRROCUMULUS and CIRROSTRATUS—High clouds; cirrocumulus has small distinct elements like grains, ripples or flakes; elements can be covered by width of finger at arm's length; "mackerel sky." Cirrostratus is milky white sheet; moon or sun haloes; distinct shadows still cast by sun.

CUMULUS—A low or middle cloud; cotton puff or cauliflower shaped. Clouds usually detached from each other; may grow to become thunderhead under proper conditions.

CUMULONIMBUS (thunderhead)—A low or middle cloud. Very massive and dense with great vertical extent around 25,000 to 30,000 feet in western Washington. Top may appear as a fibrous anvil. Showers of rain, snow or hail along with lightning.

Figure I-2 Pictures from NWS Cloud Code Chart with author's notes.

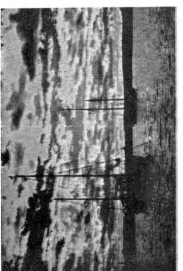

STRATOCUMULUS—A low cloud; gray and whitish in patches, sheets or layers appearing non-fibrous with cloud elements in rounded masses or rolls like loaves of French bread laid side by side. There may be open spaces between elements or they may be joined.

NIMBOSTRATUS (picture not shown)—A low cloud, extensive, dark gray with very diffuse base. Rain, snow or ice pellets fall continuously. Base often hidden by ragged clouds that change shape rapidly. This is the typical rain cloud observed in western Washington. Often accompanied by strong and blustery winds.

STRATUS—A low cloud; gray with flat base; often low enough to obscure tops of buildings and peaks of low hills. Typical cloud during summer in western Washington. Only light precipitation may occur, such as drizzle.

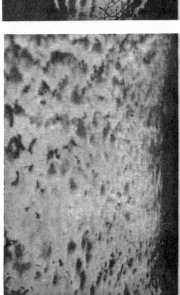

ALTOCUMULUS—A middle cloud; white and/or gray patch, sheet or layer of cloud usually composed of definite elements, such as puffy masses or rolls. Individual cloud elements can be covered by width of three fingers held at arm's length, whereas stratocumulus elements are too large to be covered. Cloud bases are higher than most mountains in the Olympics and Cascades and often higher than Mt. Rainier.

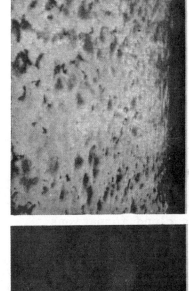

ALTOSTRATUS—A middle cloud; gray or bluish smooth cloud sheet. Sun appears as if shining through ground glass as in picture above. Cloud base higher than most mountains in Olympics and Cascades and often higher than Mt. Rainier. Shadows are not cast by sun. Rain, snow or ice pellets may fall.

Figure I-3 Cloud pictures continued.

APPENDIX J

ADDITIONAL READING

William P. Crawford	<u>Mariner's Weather</u> (W. W. Norton Co, New York, 1978).
W. J. Kotsch	<u>Weather for the Mariner</u> (U.S. Naval Institute Press, Annapolis, 1977).
Jim McCollam	<u>The Yachtsman's Weather Manual</u> (Dodd, Mead, & Co, New York, 1973).
Dag Pike	<u>Power Boats in Rough Seas</u> (International Marine Pub. Co, Camden, Maine, 1974).
Frank Robb	<u>Handling Small Boats in Heavy Weather</u> (Quadrangle - The N.Y. Times Book Co, New York, 1965).
M. J. Schroeder and C. C. Buck	<u>Fire Weather - Agriculture Handbook</u> <u>360</u> (U.S. Gov. Printing Office, Washington, D.C., 1970). This book discusses small scale weather as it affects forest fires, but is easy to read and has many diagrams which also apply to boating.
S. Townsend and V. Erics	<u>Boating Weather</u> (David McKay Co, New York, 1978).
Stuart Walker	<u>Wind and Strategy</u> (W. W. Norton Co, New York, 1973).
Alan Watts	<u>Instant Weather Forecasting</u> (Dodd, Mead, & Co, New York, 1968). An excellent reference for cloud pictures and the weather to expect with various cloud formations.
Donald A. Whelpley	<u>Weather, Water, and Boating</u> (Cornell Maritime Press, Cambridge, Maryland, 1961).

FOR LATEST RADIO BROADCAST SCHEDULES FOR WEATHER FORECASTS REFER TO:

<u>Worldwide Marine Weather Broadcasts</u> (Updated annually). U.S. Dept. of Commerce/NOAA and Dept. of the Navy. Available from U.S. Government Printing Office.

<u>Radio Aids to Marine Navigation -- Pacific</u> (Updated March and Sept.) Canadian Coast Guard Telecommunications and Electronics Branch. Available by subscription from: Printing and Publishing, Supply and Services Canada, Hull, Quebec, Canada K1A 0S9. Publication is only for Canadian stations and also lists times that various locations broadcast actual weather observations.)